Cities, Disaster Risk and Adaptation

Worldwide, disasters and climate change pose a serious risk to sustainable urban development, resulting in escalating human and economic costs. Consequently, city authorities and other urban actors face the challenge of integrating risk reduction and adaptation strategies into their work. However, related knowledge and expertise are still scarce and fragmented.

Cities, Disaster Risk and Adaptation explores ways in which resilient cities can be 'built' and sustainable urban transformation achieved. The book provides a comprehensive understanding of urban risk reduction and adaptation planning, exploring key theoretical concepts and analysing the complex interrelations between cities, disasters and climate change. Furthermore, it provides an overview of current risk reduction and adaptation approaches taken by both city authorities and city dwellers from diverse contexts in low-, middle- and high-income nations. Finally, the book offers a planning framework for reducing and adapting to risk in urban areas by expanding on pre-existing positive actions and addressing current shortfalls in theory and practice. The importance of a distributed urban governance system, in which institutions' and citizens' adaptive capacities can support and complement each other, is highlighted.

The book takes a holistic approach; it integrates perspectives and practice from risk reduction and climate change adaptation based on a specific urban viewpoint. The text is richly supplemented with boxed case studies written by renowned academics and practitioners in the field and 'test yourself' scenarios that integrate theory into practice. Each chapter contains learning objectives, end of chapter questions, suggested further reading and web resources, as well as a wealth of tables and figures. This book is essential reading for undergraduate and postgraduate students of geography, urban studies and planning, architecture, environmental studies, international development, sociology and sustainability studies.

Christine Wamsler is Associate Professor at Lund University Centre for Sustainability Studies (LUCSUS), Associate of Lund University Centre for Risk Assessment and Management (LUCRAM), Sweden, and Honorary Fellow at the Global Urban Research Centre (GURC) and the Institute for Development Policy and Management (IDPM) of the University of Manchester, UK. She is also part of the personnel pool of risk reduction experts of the Swedish Civil Contingencies Agency (MSB). In addition, Christine has been working as a consultant for different organizations, such as the International Institute for Applied Systems Analysis (IIASA), the German Society for International Cooperation (GIZ), the German, Austrian and Belgium Red Cross, the Swedish International Development Cooperation Agency (Sida), the Swedish Organization for Individual Relief (SOIR/IM) and various local NGOs. Places where she has worked and conducted research include Brazil, Chile, Colombia, El Salvador, Germany, Guatemala, India, Kosovo, Mexico, Peru, the Philippines, Sweden, Tanzania, Togo and the UK.

Routledge Critical Introductions to Urbanism and the City

Edited by Malcolm Miles, University of Plymouth,
UK and John Rennie Short, University of Maryland, USA

International Advisory Board:

Franco Bianchini   Jane Rendell
Kim Dovey   Saskia Sassen
Stephen Graham   David Sibley
Tim Hall   Erik Swyngedouw
Phil Hubbard   Elizabeth Wilson
Peter Marcuse

The series is designed to allow undergraduate readers to make sense of, and find a critical way into, urbanism. It will:

- cover a broad range of themes
- introduce key ideas and sources
- allow the author to articulate her/his own position
- introduce complex arguments clearly and accessibly
- bridge disciplines, and theory and practice
- be affordable and well designed.

The series covers social, political, economic, cultural and spatial concerns. It will appeal to students in architecture, cultural studies, geography, popular culture, sociology, urban studies and urban planning. It will be trans-disciplinary. Firmly situated in the present, it also introduces material from the cities of modernity and post-modernity.

Published:

Cities and Consumption – Mark Jayne
Cities and Cultures – Malcolm Miles
Cities and Economies – Yeong-Hyun Kim and John Rennie Short
Cities and Cinema – Barbara Mennel
Cities and Gender – Helen Jarvis with Paula Kantor and Jonathan Cloke
Cities and Design – Paul L. Knox
Cities, Politics and Power – Simon Parker
Cities and Sexualities – Phil Hubbard
Children, Youth and the City – Kathrin Hörschelmann and Lorraine van Blerk
Cities and Photography – Jane Tormey
Cities and Climate Change – Harriet Bulkeley
Cities and Nature, Second Edition – Lisa Benton-Short and John Rennie Short
Cities, Disaster Risk and Adaptation – Christine Wamsler

Forthcoming:

Cities and the Cultural Economy – Tom Hutton

Cities, Disaster Risk and Adaptation

Christine Wamsler

Routledge
Taylor & Francis Group

LONDON AND NEW YORK

First published 2014
by Routledge
2 Park Square, Milton Park, Abingdon, Oxon OX14 4RN

and by Routledge
711 Third Avenue, New York, NY 10017

Routledge is an imprint of the Taylor & Francis Group, an informa business

© 2014 Christine Wamsler

The right of Christine Wamsler to be identified as author of this work has been asserted by her in accordance with sections 77 and 78 of the Copyright, Designs and Patents Act 1988.

British Library Cataloguing in Publication Data
A catalogue record for this book is available from the British Library

Library of Congress Cataloging in Publication Data
A catalog record for this book has been requested

ISBN: 978-0-415-59102-7 (hbk)
ISBN: 978-0-415-59103-4 (pbk)
ISBN: 978-0-203-48677-1 (ebk)

Typeset in Times New Roman and Futura
by Sunrise Setting Ltd, Paignton, UK

Printed and bound in the United States of America by
Edwards Brothers Malloy

To my family

Contents

Boxes

Figures

Tables

Test yourself scenarios

Acknowledgements

The topic presented in this book is anything but theoretical or distant. It is dreadfully real and a daily concern in many people's life. During my work in disaster-prone areas around the world, many families and households living at risk have shared with me moments and insights worth a lifetime's experience, and are the 'driving force' behind this book.

This book is the result of numerous years of research and teaching the subjects of risk reduction, climate change adaptation, disaster recovery and sustainable urban development to undergraduates and graduates at Lund University, Sweden and the University of Manchester, UK. In the process of teaching these topics, the students were invaluable in asking inspiring questions, and in providing new ideas and sources of material in their course assignments and research projects. The constant discussions with my current and former colleagues at Lund University Centre for Sustainability Studies (LUCSUS), Lund University Centre for Risk Assessment and Management (LUCRAM) and the Global Urban Research Centre (GURC) have surely also contributed to this book. Consultancy work, close cooperation with other professionals and hands-on practice have complemented my academic work. This is crucial because urban risk reduction and adaptation are fields of activity where academic research and practice can, and must, through close cooperation and interaction, complement each other in order to develop sustainable solutions.

Many thanks go to the Training Regions Research Centre, the United Nations Office for Disaster Risk Reduction (UNISDR) and the various individuals who supported and contributed to this book in different ways (in alphabetical order):

- David Alexander, Professor at the Institute for Risk and Disaster Management, University College London, UK
- Per Becker, Research Manager at the Training Regions Research Centre and Associate Professor at LUCRAM, Sweden

- Terry Cannon, Research Fellow at the Institute of Development Studies (IDS), UK
- Annika Carlsson-Kanyama, Research Director at the Swedish Defence Research Agency, Sweden
- Ian Davis, Senior Professor at Lund University, Sweden and Visiting Professor at Oxford Brookes University, UK and Kyoto University, Japan
- Willemien Faling, Senior Researcher at the Council for Scientific and Industrial Research (CSIR), Built Environment, Pretoria, South Africa
- Maureen Fordham, Principal Lecturer at Northumbria University, UK
- Mohamed Hamza, Professor at the University of Copenhagen, Denmark and Senior Associate of the Global Climate Adaptation Partnership, UK
- Esteban Leon, Risk Reduction and Recovery Coordinator at the United Nations Human Settlements Programme (UNHABITAT), Geneva
- Ingrid Molander from Botkyrka Municipality, Sweden
- Helena Molin Valdes, Deputy Director of the UNISDR, Geneva, Switzerland
- Mark Pelling, Professor at King's College, London, UK
- Dan Smith, Secretary General at International Alert, UK
- Dewald Van Niekerk, Director at the African Centre for Disaster Studies, South Africa
- Janani Vivekananda, Climate Change and Conflict Adviser at International Alert, UK
- Ben Wisner, Visiting Fellow at Aon-Benfield UCL Hazard Research Centre, University College London, UK

Particularly, a huge thank you to my research assistant Ebba Brink, whose work and constant support was indispensable in finalizing this book.

And finally, the answer to Jonas' and Filippa's question posed every day during the past months:

Yes, my angels! The book is closed, and I am done. The pages full of tasks begun. A little hope, a little fear, along with dreams, are written here.

Introduction

1 Setting the scene

1.1 Rationale and scope

Disasters and other climate change[1] impacts are among today's most serious risks to sustainable urban development and an increasing concern for city authorities and planners. Worldwide, the number of disasters has almost quadrupled during the past 30 years (UNISDR 2012a) and there is widespread consensus that urban disasters are increasing exponentially,[2] resulting in escalating human and economic losses (Box 1.1). Earthquakes destroy whole cities. Flooding or landslides damage or wash away homes and infrastructure. Jobs and vital public services are lost. Heatwaves and water scarcity compromise people's health, reduce productivity, constrain the functionality of infrastructure and place cities in competition for water.

Whilst historically cities have been – and often still are – perceived as places of refuge and as buffers against environmental change, today they are better described as hotspots of disasters and risk. Recent decades have been characterized by changing climate conditions[3] and a rapid succession of major urban disasters, and even more city dwellers have lost their lives in the many small-scale disasters.[4] Risk is in fact becoming increasingly urbanized. Whilst it is the low- and middle-income nations that bear the highest burden in terms of the human lives and proportion of gross domestic product lost as a result of disasters, high-income nations are also increasingly affected.[5] Cases in point are the 2012 heatwaves in New York City, Philadelphia and Washington in the United States, the 2011 earthquake and tsunami in Japan and the 2003 European heatwave. This last resulted in more than 52,000 deaths (Larsen 2006). The tragedy was not simply the outcome of a 'natural' disaster, but the outcome of high temperatures mediated through a complex set of local conditions that made urban citizens particularly vulnerable. The modification of the land surface with materials that effectively retain heat, inadequate house constructions, lack of social networks, reduced numbers

BOX 1.1

Urban disasters and risk

Words are important, slippery, relational things. We need to be able to make clear what we mean when we use the terms disaster, risk and city – three words in the title of this book. What is it all about?

In simple terms, **disasters** are the result of an interaction between so-called natural hazards and vulnerable conditions. Hazards such as floods, earthquakes and windstorms thus do not cause disasters on their own. It is only when they are combined with vulnerable conditions, such as people or systems susceptible to the damaging effects of these hazardous events, that disasters occur, i.e. 'a serious disruption of the functioning of a community or a society involving widespread human, material, economic or environmental losses and impacts, which exceeds the ability of the affected community or society to cope using its own resources' (UNISDR 2009: 9). Disaster risk is thus the probability or likelihood that such a serious disruption occurs. The given disaster definition makes reference to different scales of impact (i.e. communities and societies) and consequently includes large-scale disasters as well as everyday small-scale disasters. Their occurrence can be related to both climatic extremes and variability.

Disasters that occur in **cities**, i.e. in an urban and not a rural context, can be called urban disasters. There is no commonly accepted definition of what is meant by the terms 'urban' and 'city'.[a] Today, however, the term 'urban' is generally viewed within the perspective of a **rural–urban continuum** spanning villages, small towns, secondary (or medium-sized) cities, metropolitan areas and megacities. Nevertheless, for practical reasons the words 'urban' and 'city' will be used interchangeably in this book.

Urban disasters are unique in the sense that they occur in an environment that has adapted to absorb large populations and services leading to specific urban characteristics in respect of (a) scale, (b) densities, (c) inhabitants' livelihood strategies, (d) economic systems and resource availability, (e) governance systems, (f) public expectations, (g) settlement structures and form, (h) likelihood for compound and complex disasters, and (i) potential for secondary impacts on surrounding rural areas and regions (see Chapter 2).

---

[a] Many countries define as cities or urban centres all settlements with populations above a threshold, for instance 1,000, 5,000 or 20,000 inhabitants, influencing the proportion of the population said to live in urban areas.

of health staff during summer periods and no contingency plans for heat-waves are only some of the contributing risk factors. Most victims were elderly people who were trapped inside their homes, living alone in neighbourhoods that lacked a sense of community and where there was a perception of danger in the streets (Dorell 2012; IPCC 2007a).

Two points can be taken from this story. First, the impact of hazards, and thus the likelihood of disaster occurrence, is strongly dependent on urban planning[6] practice and related governance processes. Second, in contrast to the notion of urban centres as places of refuge, cities actually generate hazards, both directly and indirectly – in this case, directly through increased temperatures caused by urban heat islands and indirectly because climate change and its related impacts are a result of greenhouse gas emissions, to which cities are said to be the main contributors (see Section 3.4). Consequently, there are no such things as 'natural' disasters in cities, and urban planning is capable not only of counteracting disasters and climate change impacts, but also of strongly reinforcing them (Boxes 1.1 and 1.2).

The fact that disasters are not 'natural', but rather a socio-economic construct, is also illustrated by the many poor or underprivileged who are particularly at risk, in both the southern and the northern hemisphere. They often settle on marginal land, have substandard housing and infrastructure, do not receive early warnings or, as in the case of Hurricane Katrina in New Orleans, are not able to leave the city. While poor and marginal living conditions reinforce people's vulnerability to hazards, disasters make their already precarious conditions worse, creating a vicious circle of increasing risk that all too often results in poverty traps.

As a consequence of the described situation, city authorities and planners are increasingly facing the challenge of finding ways to include risk reduction and adaptation strategies in their work, although related knowledge and competence is still scarce and fragmented. During the past two decades, increasing attention has been given to the field of disaster risk reduction, at first mainly within the context of disaster response (DFID 2004; Twigg 2004). It is only in recent years that more and more consideration has also been given to the need to address disasters and risk through development work so as to attain sustainable risk reduction *and* climate change adaptation (Box 1.3). A range of different policy documents allude to this need, including the Millennium Declaration and the Hyogo Framework for Action 2005–2015, which urges governments to address the issue of disaster risk in different sector work (UNISDR 2005).

To put cities in the spotlight and bring city authorities and planners 'on board', the 2010–2015 World Disaster Reduction Campaign on 'Making Cities

BOX 1.2

Urban disasters and climate change

Disasters are triggered by so-called natural hazards, which include both climatic and non-climatic hazards. Climatic hazards are, for instance, precipitation, floods, windstorms, droughts, fires, heat- and coldwaves, sea-level rise (water surges) and landslides. These hazards can arise from, and materialize in the form of, both climatic extremes and variability. Non-climatic hazards include, for instance, earthquakes and volcanic eruptions. Disaster risk refers to risk related to climatic and non-climatic hazards, whilst climate risk only refers to risk related to climatic hazards.

The terms weather-related or non-weather-related hazards are commonly used as synonyms for climatic and non-climatic hazards, although climate is not the same as weather. The average pattern of weather is called climate and usually stays pretty much the same for centuries if not exposed to external influences. However, the Earth is not being left alone. According to the Intergovernmental Panel on Climate Change (IPCC), concentrations of greenhouse gases, including carbon dioxide (CO_2), methane (CH_4), chlorofluorocarbons (CFCs) and nitrous oxide (N_2O), have increased dramatically since 1750, primarily as a result of human activities, leading to climate change (IPCC 2007c; Bulkeley and Betsill 2003).

Climate change is closely related to 'global warming', which refers to an increase in mean annual surface temperature of the Earth's atmosphere and in turn results in changing climate conditions in terms of both climatic extremes and variability. Importantly, these changing climate conditions can aggravate both the existing hazards and the present vulnerable conditions (see Box 1.1), thus considerably increasing risk and disaster occurrence. Feedback loops between disasters and climate change are further reinforced by the fact that urban areas already at risk from disasters are those most likely to be affected by climate change in the future (see Chapter 3).

Resilient' was created. It supports the five priorities of the Hyogo Framework for Action by linking them to the urban context with the aim being to:

1. Make disaster risk reduction a priority in urban practices
2. Create a knowledge base on urban risks
3. Build understanding and awareness of urban risks at all levels
4. Reduce urban risks, and
5. Prepare cities, making them ready to act (UNISDR 2010a).

Further international key policy documents relevant to urban risk reduction *and* adaptation are, amongst others:

- Agenda 21, a product of the United Nations Conference on Environment and Development (UNCED) held in Rio de Janeiro in 1992 and informally known as the Earth Summit. This is a non-binding action plan to support sustainable development, affirmed and modified at subsequent UN conferences.[7]
- The United Framework Convention on Climate Change (UNFCCC), an international environmental treaty, also produced at the Earth Summit in 1992. The treaty provides for updates (called 'protocols') that would set mandatory limits on greenhouse gas emissions. The principal update is the Kyoto Protocol.
- The Kyoto Protocol, initially adopted in 1997 in Kyoto, Japan, entered into force in February 2005.
- The Cancun Adaptation Framework, established at the 2010 Climate Change Conference in Cancun, Mexico in 2010 (i.e. the 16th session of the Conference of the Parties [COP 16] to the UNFCCC and the 6th session of the Conference of the Parties to the Kyoto Protocol).
- The Global Framework for Climate Services (GFCS), established during the Third Climate Change Conference in September 2009 in Geneva to enable better adaptation to climate change through the development and incorporation of science-based climate information and prediction into planning, policy and practice.

Finally, the Rio+20 world conference in June 2012 included for the first time the issues of sustainable cities *and* disaster risk reduction as (two out of seven) priority areas in the global sustainable development agenda.[8] The final report of the conference states:

> We underline the importance of considering disaster risk reduction, resilience and climate risks in urban planning.
>
> (United Nations 2012: 26)

Apart from the international policy documents, there are relevant legislation and policy directives at regional levels. An example is the European Strategic Environmental Assessment (SEA) Directive, which legally obliges planners to consider climate change in urban development planning.[9]

Despite the advances at the international and regional policy level, city authorities, aid organizations and planners are still struggling to effectively tackle disaster risk through their everyday work. This is a result of, amongst

---

BOX 1.3

Urban risk reduction and climate change adaptation

Risk reduction and, even more, climate change adaptation are still relatively new areas of knowledge. Whilst they have in the main developed independently, they share the aim to reduce the occurrence and impacts of climate-related disasters and associated risk, which includes climatic extremes and variability (see Chapter 2). Risk reduction and adaptation can, therefore, not be meaningfully conducted without considering climate variability. In contrast to climate change adaptation, risk reduction also addresses non-climate-related risk and disasters. Importantly, both areas are cross-cutting issues. This means that both risk reduction and adaptation need to be integrated into urban planning and all kinds of sector work – implemented before, during and in the aftermath of disaster occurrence (i.e. the context of development, disaster response and disaster recovery). This book presents a holistic approach throughout in integrating perspectives and practice from risk reduction and climate change adaptation based on a specific urban viewpoint.

---

other things, the lack of (access to) adequate knowledge and tools that are relevant and applicable at a city and local level (e.g. Carmin *et al.* 2012; Pelling 2007; SKL 2011; UNISDR 2010a, 2010c, 2012c; Wamsler 2009a). Those working in urban planning still have a tendency to think about risk reduction and adaptation in a purely physical way,[10] ensnaring themselves in constructive (high-)tech discussions and discourses. Such discussions, however, tackle only a small part of the necessary and possible measures, and too often ignore the root causes of vulnerability.[11] As a consequence, while city authorities and planners have the responsibility for developing secure and sustainable cities, they often contribute to the increase in risk and disasters. A consensus thus has emerged that 'something' needs to be done, whilst questions as to what, how and when remain to a large extent unanswered.

Against this background, *Cities, disaster risk and adaptation* addresses the urgent need to re-evaluate current city planning for risk reduction and adaptation, which is required to achieve sustainable urban development. In order to do so, it brings together complex disaster risk and adaptation literatures, with a specific urban focus, in a coherent, thorough and holistic analysis. With disasters being a product of past developments, responding and adapting effectively to disasters and associated risk is inherently complex. This book challenges the dominant solutions – physical planning and grand engineering projects – that provide security for some but exclude many more. It shows that the widely recognized need to incorporate better knowledge into urban planning for safer houses and infrastructure is just one of many issues that planners must address. In addition, it shows how sustainable urban development practices can flow from high- to low-income nations as well as from low- to high-income nations. In many poor countries, the need to compensate for inadequate urban risk governance structures and reimagine the city provides a rich context for new urban practice.[12]

1.2 Overall aim

The overall aim of this book is to enable students and planners working for city authorities and other urban actors[13] to understand how urban communities can become more disaster resilient and able to counteract increasing disasters and climate change impacts – rather than inadvertently reinforcing them (Box 1.4). More specifically, it aims to:

- Analyse the complex interrelations between disasters, risk and urban development processes
- Provide an overview of current planning practice for risk reduction and adaptation
- Provide a conceptual and operational understanding of how to (better) integrate risk reduction and adaptation into urban planning practice
- Explore in this context the roles that citizens' local efforts, urban institutions[14] and related governance systems (can) play in achieving sustainable urban transformation.

After reading this book, the reader will have obtained and be able to demonstrate:

- A holistic view and systems perspective of the main processes of urban disaster management, risk reduction, climate change adaptation, disaster response and recovery
- A critical understanding of different concepts and issues central to the understanding of risk reduction and adaptation in urban contexts
- Knowledge on concrete measures and operational tools for integrating risk reduction and adaptation into city planning,[15] related urban governance structures and capacities.

BOX 1.4

Resilient cities

From a disaster risk perspective, resilience refers to '[t]he ability of a system, community or society exposed to hazards to resist, absorb, accommodate to and recover from the effects of a hazard in a timely and efficient manner . . .' (UNISDR 2009). Resilience is thus an attribute of a system's behaviour, indicating how well the system performs. On this basis, the ideal of urban resilience is here understood as cities that can easily withstand, cope with and overcome disasters, including climate- and non-climate-related, small- and large-scale disasters. The aim is to restore the historical function of cities as places where citizens can find safety and protection from disasters and environmental change. It is hoped that this book will help this vision become reality.

Imagine cities that effortlessly cope with hurricanes, fires or floods, that reinforce themselves and seal cracks of their own accord, and buildings that elevate themselves during flooding. Imagine cities that provide information systems that warn when a hurricane is approaching or when houses are overburdened and may be liable to imminent collapse. Such cities would secure the livelihood of all their inhabitants, empowering them to cope and deal with natural threats. As with a living organism, such cities would adjust their social, political and economic systems in such a rapid way that they can account for damage, effect repairs, learn from experience, transform and retire – urbanely – once they can no longer fulfil their protective and defensible function (adapted from Wamsler 2006a). (See Chapter 2, Boxes 2.2 and 2.3.)

Cities, disaster risk and adaptation thus supports and contributes to the priority areas of the Hyogo Framework for Action 2005–2015 and the goals of the 2010–2015 World Disaster Reduction Campaign 'Making Cities Resilient'. Whilst the book's focus is on cities and urban planning, its content has much wider applications. It is a useful source for both undergraduate and postgraduate students across the disciplines of environmental studies, geography, international development, sociology, sustainability studies, urban studies and planning. Its institutional target group includes, amongst others, government officials and practitioners, urban planners, and staff from non-governmental and community-based organizations and environmental consultancies from both developed and developing contexts.

1.3 Outline

Chapter 1 sets the scene by defining the scope and aim of the book. On this basis, the book is structured into three parts:

1. Theory
2. Practice
3. Moving forward.

Part 1 comprises two chapters. Chapter 2 provides an overview of key concepts and terms central to the understanding of urban disaster risk management, risk reduction and adaptation (such as risk, hazard, vulnerability, adaptive capacity, mitigation, preparedness, mainstreaming, resilience and sustainable transformation) and the recent evolution of thinking on these concepts. Chapter 3 explores the reciprocal linkages between disasters and cities, analysing in detail how the urban fabric with its characteristic urban features influences both climate- and non-climate-related risk.

Part 2 deals with the current practice of urban planning for risk reduction and adaptation in high-, middle- and low-income nations. Chapter 4 presents an overview of the prevalent measures and strategies implemented by city authorities. Chapter 5 focuses on city dwellers' local initiatives to reduce and adapt to increasing disasters and climate change.

Part 3 compares the theoretical and practical approaches presented in Parts 1 and 2 and, on this basis, elaborates on the notion of resilient cities and sustainable urban transformation and how it could be (better) achieved in the future (see Box 1.4). In Chapter 6 existing differences, gaps and related challenges are discussed, together with ways in which these could be bridged. Concrete strategies are provided whereby city authorities and planners could better fulfil their responsibility for creating resilient cities. Chapter 7 briefly revisits some of the main themes of the book. It provides succinct conclusions about the arguments and findings presented in the previous chapters. The consequences for how we conceive of urban planning and the city in an era of increasing disasters and climate change are discussed.

1.4 Format and style

The book is written in a textbook or lecture style to provide the reader with a comprehensive understanding of urban disaster risk reduction and climate change adaptation, without getting 'lost' in theoretical discourses with little practical relevance. Each chapter has a similar format, starting with an out-line of the key learning objectives addressed in each chapter. The following text is structured with the help of 'guiding questions' and is accompanied by various figures, tables and boxes to provide additional input and effective illustrative examples to support, or highlight, some of the key issues. Several internationally recognized experts contributed to this, including (in alpha-betical order): David Alexander, University College London, UK; Per Becker, Training Regions Research Centre and Lund University, Sweden; Terry Cannon, Institute of Development Studies (IDS), UK; Annika Carlsson-Kanyama, Swedish Defence Research Agency; Ian Davis, Lund University, Sweden, Oxford Brookes University, UK and Kyoto University, Japan; Willemien Faling, Council for Scientific and Industrial Research (CSIR), South Africa; Maureen Fordham, Northumbria University, UK; Mohamed Hamza, Copenhagen University, Denmark and the Global Climate Adaptation Partnership; Esteban Leon, United Nations Human Settlements Programme (UNHABITAT), Geneva; Ingrid Molander, Botkyrka Munici-pality, Sweden; Helena Molin Valdes, United Nations Office for Disaster Risk Reduction (UNISDR), Geneva; Mark Pelling, King's College, London, UK; Dan Smith, International Alert, UK; Dewald Van Niekerk, African Centre for Disaster Studies, South Africa; Janani Vivekananda, Interna-tional Alert, UK; and Ben Wisner, University College London, UK.

With the aim of stimulating wider interest, engagement and critical thinking, every chapter includes a summary for action taking to highlight some key aspects and ends with (a) suggested learning activities (questions and 'test yourself' scenarios that the reader should be able to discuss, answer or solve after having studied each chapter); (b) a list of key references for further reading; and (c) a list of relevant websites. The suggested learning activities can be used for self-study or by academics/teachers in the context of training courses. A complete bibliography and index appear at the end of the book.

In sum, this book provides an objective-oriented and well-guided learning base that encourages the reader to take an active role by answering ques-tions, working on case studies, deepening certain areas of knowledge and exploring real-life examples and challenges of urban risk reduction and cli-mate change adaptation when following suggested references and online resources.

Part 1
Theoretical framework

2 Sorting out the conceptual 'jungle' associated with urban risk reduction and adaptation

Learning objectives

- To gain an overview of key concepts and terms central to the understanding of urban disaster risk management, risk reduction and climate change adaptation

  ○ To identify what constitutes urban risk in conceptual terms
  ○ To identify the conceptual characteristics of disaster-resilient cities

- To become acquainted with a conceptual and operational framework for mainstreaming risk reduction and adaptation into urban planning practice

The worldwide increase in urban disasters makes the constant struggle and failure of city authorities and other urban actors all too visible. Improved knowledge and capacity on conceptual and operational frameworks to better address this situation are crucial. But what are the concepts and terms city authorities and planners need to be familiar with when dealing with disasters and planning for risk reduction and adaptation? Unfortunately, there is no short answer to this question. Even the most simplistic and popular documentaries and publications on these issues confront the audience with an overwhelmingly large number of different concepts. This conceptual 'jungle' includes terms such as risk, vulnerability, hazard, prevention, mitigation, preparedness, reconstruction, resilience, sustainable transformation and many more (Figure 2.1). Too often these concepts are used without further explanation. Whilst there are no universally accepted definitions of these terms, being able to interrelate them in a coherent way is crucial in order, ultimately, to 'construct' a framework that can guide the comprehensive management of increasing urban disasters and changing risk patterns in

Figure 2.1 Word cloud presenting the conceptual 'jungle' associated with urban risk reduction and adaptation. Source: graphic created in Wordle.net.

practice. So how can we find our way through the existing conceptual jungle and make sense of its many different terms?

Different answers are possible and depend on the definition of each concept. The answer presented in the following sections is considered the most suitable for city authorities and other urban actors because it provides a coherent conceptual framework for reducing and adapting to risk in urban areas, which furthermore can be easily operationalized.[1]

In simple terms, the key concepts and their interrelation can be summarized as follows. There are three main **processes** of disaster risk management, namely (1) disaster response, (2) disaster recovery and (3) disaster risk reduction (see Table 2.1, last column, bold terms). The third process, disaster risk reduction, is closely related to climate change adaptation. Both processes share the aim of increasing disaster resilience by reducing the number and impacts of climate-related disasters and associated risk, which includes climatic extremes and variability (see Boxes 1.2–1.3). The concepts of disaster, disaster risk and resilience can thus be seen as the **conceptual basis** that underlies urban risk reduction and adaptation (see Boxes 1.1 and 1.4 and Table 2.1, first column, bold terms). Risk reduction and adaptation include five main **activities**, namely (1) hazard reduction and avoidance, (2) vulnerability reduction, (3) preparedness for response, (4) preparedness for recovery and (5) risk assessment (see Table 2.1, second column, bold terms). If people at risk take measures on their own to reduce or adapt to

Table 2.1 Categorization of key terms and concepts central to the understanding of disaster risk management, risk reduction and climate change adaptation

Underlying basic concepts	Measures and activities	Processes and working fields
Disaster Complex emergency Conflict (Crisis) **Disaster risk** Climate risk Vulnerability Hazard Climatic extreme Climatic variability Capacity Adaptive capacity Coping capacity **Resilience and transformation** Disaster resilience Sustainable (urban) transformation	**Hazard reduction and avoidance** Prevention Hazard reduction Hazard avoidance Climate change mitigation **Vulnerability reduction** Disaster mitigation **Preparedness for response** Contingency planning Evacuation planning **Preparedness for recovery** Risk financing Risk transfer Risk sharing **Risk assessment** Risk analysis Risk evaluation **Coping strategies** Individual practice for risk reduction Private adaptation Autonomous adaptation Adaptive behaviour **Mainstreaming** Different strategies (Boxes 2.8–2.10)	**Disaster response** Response Humanitarian assistance Emergency management (Disaster) relief (Disaster) rescue **Disaster recovery** Recovery Reconstruction Rehabilitation Early recovery **Disaster risk reduction** Risk reduction Climate change adaptation **Development work** Technical cooperation Technical assistance

disasters and risk, they are generally called coping strategies (see Table 2.1, second column, bold terms). Risk assessment provides the knowledge base for the identification, design and implementation of the risk reduction and adaptation **measures** (1)–(4). With both risk reduction and adaptation being so-called cross-cutting or mainstreaming issues, these activities need to be integrated into the context of (a) disaster response, (b) disaster recovery and (c) development work (see Table 2.1, second and third column, bold terms).

The conceptual framework just described shows that it is helpful to be able to differentiate between processes, measures and activities associated with urban risk reduction and adaptation, and the related underlying concepts (Table 2.1). As a first step, these aspects can be used to make sense of the many different terms presented in Figure 2.1:

1. The **basic concepts** that underlie both the processes and related measures of risk reduction and adaptation. These not only refer to the key concepts of disaster, disaster risk and disaster resilience mentioned previously, but also include other concepts such as complex emergencies, conflict, crisis, climate risk, vulnerability, hazard, climatic extremes and variability, coping capacity and adaptive capacity (see Table 2.1, first column).
2. Existing **measures and activities** that can be employed within the context of risk reduction and adaptation. Apart from the previously mentioned concepts of hazard reduction and avoidance, vulnerability reduction, preparedness for response, preparedness for recovery, risk assessment, coping strategies and mainstreaming, this category also includes other concepts such as prevention, mitigation, contingency and evacuation planning, risk financing, risk transfer and sharing, risk analysis and evaluation, private or autonomous adaptation and adaptive behaviour (see Table 2.1, second column).
3. Existing **processes** or working fields of disaster risk management and, more specifically, urban risk reduction and adaptation. Apart from the previously mentioned processes of disaster response, disaster recovery and development work, this category also includes humanitarian assistance, emergency management, disaster relief and rescue, reconstruction, rehabilitation, early recovery, technical cooperation and technical assistance (see Table 2.1, third column).

In order to find our way through the conceptual jungle presented in Figure 2.1, we now need to have a more in-depth look at the terms listed within each category. They can be grouped under umbrella terms. In Table 2.1, these umbrella terms are written in bold and include the other terms listed, which are either subcategories or synonyms. All concepts and their interrelations are described in detail in the following text, starting with the underlying basic concepts, followed by the related measures and processes.

2.1 Basic concepts

A disaster can be defined as 'a serious disruption of the functioning of a community or a society involving widespread human, material, economic or environmental losses and impacts, which exceeds the ability of the affected community or society to cope using its own resources' (UNISDR 2009: 9; Box 1.1). Disaster risk is the probability or likelihood that such a serious disruption occurs, which is determined by the interaction between so-called natural hazards (*H*) and vulnerable conditions (*V*). Disaster risk is thus conventionally expressed in the following pseudo-equation:

$$R = H \cdot V \qquad \text{(Equation 1)}$$

where *R* stands for disaster risk, *H* for hazard(s) and *V* for vulnerability.

This pseudo-equation clearly illustrates the fact that hazards do not cause disasters on their own. They are not disasters per se. Rain and floods might, in fact, be a most welcome event for people dependent on shrimp farming in Bangladesh and an earthquake far away from inhabited areas might be of little concern to us. It is only when hazards are combined with vulnerable conditions, such as people and related systems susceptible to the damaging effects of these hazardous events, that risk might become unmanageable. It is only then that hazards 'translate' into disasters.

What are the key aspects urban actors need to remember from the described understanding of risk and disaster? It is important to recall that:

- Risk and disasters are determined by the same factors (*H* and *V*)
- Disasters are the outcome of (continuously) present conditions of risk
- An event can only be called a disaster if it (a) is triggered by the combination of a 'natural' hazard (or several hazards) and vulnerable local conditions, (b) results in a disruption of the functioning of a community or society, and (c) requires external assistance for the subsequent impacts to be adequately dealt with (Figure 2.2).

This understanding of disaster does not, however, imply that it is only about large-scale events related to climatic extremes. UNISDR's definition of disaster makes reference to different scales and levels of impact by referring to communities and societies, and thus includes both large-scale and everyday small-scale disasters. In case a specific area (or city) within a certain city (or country) is in need of external assistance to cope with a hazardous event, this event presents a disaster at city (or national) level. If it is a country that calls for assistance from other states, then we are dealing with a so-called international disaster (i.e. a disaster at the international level).

Disasters that occur in cities, or any other kind of urban environments, can be called urban disasters (see Box 1.1). They are unique in the sense

Serious disruption of community
External assistance required

Figure 2.2 The disaster concept. The illustration is based on UNISDR's definition of risk (UNISDR 2009: 25), as well as the Pressure and Release Model presented in Wisner *et al.* (2004: 49).

that they occur in a dense and highly complex, physical and non-physical environment that has adapted, formally and informally, to absorb large populations and services. The characteristic features of urban disasters will be discussed in detail in Chapter 3.

After having created a common ground for understanding disasters and associated risk, we can now have a closer look at the different risk factors. The first risk factor is H (see Equation 1). Before continuing to read, ask yourself: what is a so-called natural hazard? Why is it called 'natural' and what does it include? And why are cities (increasingly) exposed to hazards?

Disasters can be triggered by both climatic hazards and non-climatic hazards. The term disaster risk therefore refers to risk related to climatic and non-climatic hazards, whilst the term climate risk only makes reference to risk related to climatic hazards/disasters (Box 1.2). Climatic hazards are, for instance, precipitation, floods, wind- and snowstorms, droughts, fires, heat- and coldwaves, sea-level rise (water surges) and landslides (Table 2.2). They can arise from, and materialize in the form of, both climatic extremes and variability. Non-climatic hazards include, for instance, earthquakes and volcanic eruptions (Table 2.3).

Both types of hazards can be called a 'natural hazard', although this term is highly misleading because most hazards do also have a social or human trigger (see Chapter 3 and Box 1.2). Nevertheless, the term natural hazard is commonly accepted and justifiable to differentiate hazards with a (socio-)natural trigger from other so-called man-made hazards, such as civil

Table 2.2 Climate-related hazards in an urban context

Natural hazards	Urban hazards and concerns[a] Factors that influence increasing hazard exposure of urban populations (including exposure to climatic extremes and variability)
Drought (hydrometeorological hazard)	Drought is increasingly occurring in urban areas. This situation can be related to different issues, including increasing temperatures and reduced rainfall (due to climate change), the urban heat island effect and many city dwellers' lack of understanding about the effective and responsible use of water. Drought in rural areas or small urban centres can trigger migration to cities and this increases pressure on urban housing, infrastructure and basic services and their development on hazardous land.
Temperature, heat- and coldwaves (hydrometeorological hazard)	Cities are generally warmer than their surrounding areas. Urban surfaces such as buildings, asphalt and short grass heat up rapidly during the day, which can result in city temperatures up to 10 degrees Celsius higher than in nearby woodlands. Climate change can further increase some cities' exposure to high temperatures. Hot and humid cities are often also prone to strong storms (e.g. hurricanes in the Caribbean islands and southeast United States, and typhoons in southeast Asia and northeast Australia), confronting urban areas with a complex set of different hazards. Due to climate change, many cities may also be increasingly affected by strong winters or periods that are colder than usual, with consequences such as slippery streets, related accidents, increased energy demand and/or use of firewood for heating, flu and pneumonia and deaths among homeless people. The impacts of severe winters in rural areas, such as destruction of crops and livestock, can strongly affect urban areas (e.g. by causing food shortage and massive rural–urban migration).
Precipitation, flood and sea-level rise (hydrometeorological hazard)	Many urban areas are in places where they are exposed to heavy rain, situated on riverbanks or near deltas. Currently approximately 2.8 billion people – more than 40 per cent of the world's population – live in coastal cities. Urban areas generally have more rainfall compared with the surrounding areas. Due to climate change, some cities are (or will be) affected by sea-level rise and/or increase in precipitation. As regards the latter, flash floods are a growing urban hazard because drainage systems are inadequate or blocked by solid waste, concrete or compacted earth does not absorb water, open spaces have been colonized and engineering works have changed river flows. In rural areas, even small variations in rain patterns can negatively affect crops and thus impact the rural–urban food flow.

(continued)

Table 2.2 (*continued*)

Natural hazards	Urban hazards and concerns[a] Factors that influence increasing hazard exposure of urban populations (including exposure to climatic extremes and variability)
Wind- and thunderstorm (hydrometeorological hazard)	Many urban areas are exposed to strong winds (including cyclones, typhoons or hurricanes), which may increase due to climate change. A city's layout may also increase wind speeds due to the so-called 'wind tunnel effect'. Storms are often combined with heavy rain or thunder and trigger secondary hazards, such as floods and landslides as well as fires due to lightning or falling electric posts. The urban fabric itself might further attract lightning.
Fire and wildfire (hydrometeorological hazard or technological)	Urban fires stem, for instance, from uncontrolled wildfires, windstorms, earthquakes, high temperatures, industrial explosions or domestic accidents in densely built settlements. Rapid growth of cities multiplies the fire risk by, for instance, increasing building densities or introducing new building materials that are not fireproofed. Wildfires occur in rural areas and on the peripheries of cities. Uncontrolled wildfires can reach urban areas, as well as trigger migration to cities.

Sources: IOI 2012; IPCC 2007a; Kim 1992; UNHABITAT 2011; UNISDR 2010a; Wamsler 2007c; Wisner *et al.* 2004; see also Chapter 3.

Note:
[a] Examples of major cities affected by the mentioned hazards: cities where drought is becoming increasingly common are Sydney and other cities in Australia, cities in Texas and California (United States) and Seville (Spain). Cities prone to cold temperatures or severe winters are, for instance, Beijing (China), Ulan Bator (Mongolia) and Fort Yukon (Alaska). Examples of cities at risk from flooding are São Paulo (Brazil), Mumbai (India) and Lagos (Nigeria). Cities at risk from sea-level rise include Dhaka (Bangladesh), Alexandria (Egypt) and Amsterdam (Netherlands). Examples of cities prone to cyclones/hurricanes are Shanghai (China), Kolkata (India) and San Salvador (El Salvador). Cities at risk from hailstorms are, for instance, Amarillo (Texas, United States), Sydney (Australia) and Jaipur (India). Examples of cities where wildfires increasingly occur in the peripheries are Los Angeles (California, United States), Sydney (Australia) and Cape Town (South Africa).

disorder, crime, war, terrorism, violence, industrial hazards and traffic accidents. If man-made hazards (sometimes also called crises) escalate, one generally talks about conflicts (and not necessarily about disasters). Since conflicts and disasters are frequently linked through common causes and not infrequently happen at the same time, conflict and disaster risk management are in practice often combined (UNDP 2011). The prevalence of both natural and man-made hazards can lead to so-called complex disasters or emergencies. Health hazards are partly considered as natural hazards,

Table 2.3 Non-climate-related hazards in an urban context

Natural hazards	Urban hazards and concerns[a] Factors that influence increasing hazard exposure of urban populations
Earthquake (geological hazard)	Many cities lie on earthquake belts. New research suggests that earthquakes can also be triggered by the melting of permafrost, which indicates that even earthquakes may sometimes be categorized as climatic hazards.
Tsunami and water surge (geological hazard)	Many cities are located on tsunami-prone coasts. The tsunami risk is believed to be highest on the coasts of the Pacific Ocean and the Indian Ocean. Tsunamis are triggered by other hazard events, such as underwater earthquakes and landslides or land-based or submarine volcanic eruptions.
Volcanic eruption (geological hazard)	Historically, humans have been drawn to volcanoes because of the surrounding rich soils and flat terrains. Many of these settlements have turned into urban centres, with millions of people living today on volcanic flanks or in potential paths of mud or lava flows (Figure 2.3).
Landslides (geological hazard)	Many urban settlements are located on steep and unstable slopes caused by different urbanization processes (e.g. the colonization of former waste disposal sites or landfills, reduced vegetation and inadequate construction on slopes). Landslides are often secondary hazards that are triggered by earthquakes or result from different climate-related hazards (e.g. heavy rain, flooding, melting snow, rapidly decreasing water levels at the base of the slope). Under such circumstances they can thus also be considered as climatic hazards.
HIV/AIDS and other diseases (biological hazard)	HIV/AIDS and other diseases are major barriers to city dwellers' capacity to adapt to other types of hazards and to cope with related impacts. Inadequate housing and settlements, high densities and overcrowding increase urban dwellers' risk of both opportunistic infections and HIV/AIDS. To make matters worse, climate change is predicted to increase both water- and vector-borne diseases (such as cholera, dengue and malaria). Even small variations in, for example, rain patterns and temperatures can lead to changes in the breeding and habitats of disease vectors.

Sources: IPCC 2007a; Turpeinen *et al.* 2008; UNAIDS 2008; UNISDR 2010a; Wamsler 2008d; Wisner *et al.* 2004. See Chapter 3.

Note:

[a] Examples of major cities affected by the mentioned hazards: cities that lie on earthquake belts are, for instance, Tehran (Iran), Istanbul (Turkey), Bogotá (Colombia), Manila (Philippines) and Kathmandu (Nepal). Prone to tsunamis are coastal cities in, for instance, Japan, Hawaii and Peru. Cities threatened by potential volcanic eruptions are, for example, Jakarta (Indonesia), Manila (Philippines) and Mexico City (Mexico). Small-scale landslides are common in many marginal settlements. Large-scale landslides have occurred in major cities such as Rio de Janeiro (Brazil), Caracas/La Guaira (Venezuela) and Cairo (Egypt).

resulting in health disasters or epidemics (including, for instance, flu or AIDS epidemics).

Another important concept is the so-called secondary hazard. It is a hazard that has been triggered by another preceding natural hazard, such as fire caused by drought, lightning or earthquakes; diseases caused by floods; or landslides caused by rain. In many cases, the secondary hazard can be more destructive than the primary one. This is one of many reasons for not

Figure 2.3 A house buried after the 1991 eruption of Mount Pinatubo in the Philippines; next to it is the homeowners' new house (in the background), constructed to resist future mud and lava flows. Source: Christine Wamsler.

addressing climatic and non-climatic hazards in isolation and for promoting the use of a multi-hazard approach to manage increasing disasters and risk.

Tables 2.2 and 2.3 list the key aspects and concerns related to natural hazards in urban environments. The factors that influence urban populations' increased hazard exposure are presented. They clearly show that so-called natural hazards often have a socio-natural trigger and, in addition, are strongly influenced by complex urban–rural linkages. Table 2.2 further includes examples to illustrate the multifaceted interrelationship between hazards, climatic extremes and variability.

Vulnerability is the second factor influencing risk. Before continuing to read, ask yourself: what does the term vulnerability mean? How can it be defined?

In contrast to the term hazard, vulnerability is a more complex and multifaceted concept. It is not a synonym for disaster risk, although it is often used this way. In simple terms, it is the degree to which communities or societies are 'susceptible to the damaging effects of a hazard' (UNISDR 2009: 30). It thus describes the existing conditions, characteristics and circumstances of an area exposed to one or several hazards, where a highly vulnerable area is understood as being incapable of resisting their impacts. But why are some areas more vulnerable than others when facing natural hazards? Before continuing to read, ask yourself: what are the factors that can make a city more (or less) vulnerable? Or in other words: why do some cities deal more (or less) effectively with hazards than others?

The first factor that influences a city's degree of vulnerability is location-specific conditions, such as the quality of the built environment, population densities, natural resource degradation and existence of peace and security. These local conditions strongly influence the level of hazard impact, and they are independent of (and thus have to be distinguished from) any reactions by city authorities or other urban actors to deal with the hazard impacts.

The second factor that influences a city's degree of vulnerability is the functioning of people's and institutions' reactions in dealing with hazard impacts. More specifically, it is the functionality of existing mechanisms and structures for (a) disaster response and (b) disaster recovery that can make a city more or less vulnerable.

All three aspects that influence an area's degree of vulnerability (i.e. the location-specific conditions, existing response mechanisms and structures, and existing recovery mechanisms and structures) have distinct features but are closely interconnected and can all become 'visible' in physical, environmental, social, economic and political/institutional terms. They can, for instance, be apparent in the existing building fabric, compliance with laws, levels of corruption, resource availability or the institutional infrastructure.

But what does the described understanding of disasters, hazards and vulnerability mean in concrete and practical terms? Let's explain this with the help of an illustrative example. Imagine an area prone to a specific hazard, for instance an earthquake-prone city. Once an earthquake occurs, it is not only the magnitude, intensity and time span of the hazard that determine whether it will result in a disaster or not. This depends also on the city's ability to resist or cope with the earthquake, i.e. its pre-existing local conditions and the mechanisms and structures that people and institutions have at their disposal to react to the hazard impacts. In sum, there are thus four risk factors that determine whether the result will be a disaster (or 'only' a hazard that a specific community can manage on its own). These are:

1. Current and future hazards (i.e. type, magnitude, intensity, time span)
2. Current and future susceptibility of the location making it unable to withstand existing and future hazards (i.e. pre-conditions, not related to reactions after hazard occurrence)
3. The functioning of mechanisms and structures in place to respond to current and future hazards
4. The functioning of mechanisms and structures in place to aid recovery from current and future hazards.

Whilst the last three aspects listed can be grouped under the concept of vulnerability, it is crucial for urban actors to know about, and work with, these sub-components. Dividing the concept of vulnerability into these sub-components facilitates a better grasp of its meaning and complexity and makes it more operational. Knowledge of all four risk factors can, in fact, be seen as one of the most important prerequisites for city authorities, aid organizations and planners, allowing them to analyse urban risk and, on this basis, find adequate measures to reduce the identified risk in a comprehensive way.

In order to further find our way through the conceptual jungle associated with risk reduction and adaptation, it is crucial to understand how the concept of risk is linked to the other terms listed in Table 2.1 (first column), namely capacity, adaptive capacity, resilience and transformation. Before continuing to read the following section, ask yourself: how are adaptive capacity and resilience linked to the other concepts already discussed?

The concept of capacity is probably the one that creates most confusion and misunderstanding. The reason for this is the fact that it is used in two different contexts, and its meaning (or definition) changes depending on the context in which it is used. On the one hand, capacities are discussed in the context of assessing the risk of a specific area (which can be done after or before the occurrence of a hazard or disaster). On the other hand, capacities are discussed in the context of identifying potential measures to reduce or adapt to the identified risk.

In the context of assessing the risk of a specific area, the term capacity commonly refers to the existing risk reduction measures in place. In other words, it is the (already) *used capacity* of different stakeholders to reduce or adapt to risk that is visible in the form of different kinds of actions taken. This meaning of the term capacity is also reflected in the disaster definition, with disasters being the result of 'insufficient capacity or measures to reduce or cope with potential negative consequences' (UNISDR 2009: 9). Examples of *used capacity* can be the existence of early-warning systems, public awareness (campaigns), emergency shelters and the elaboration of and compliance with building codes. It includes measures at the institutional level as well as so-called local coping strategies of individuals, households and communities. Notably, the capacity, which becomes visible in the form of concrete measures, includes both functioning and sustainable as well as non-functioning and unsustainable ones.

In contrast, in the context of identifying potential risk reduction and adaptation measures, the term capacity also refers to people's and institutions' *potential* to reduce and adapt to risk. In other words, it not only includes the institutionally based measures and the local coping strategies that are already in place; it also includes existing capacities that are not (yet) used to reduce and adapt to risk. Examples of *unused* capacities can be the existence of local builders (who could be trained in disaster-resistant construction), the existence of a community centre (which could be used for training or as an emergency shelter), well-established local organizations and associations (which are interested in becoming engaged in risk reduction and adaptation), or microfinancing institutions (which could develop disaster insurance policies). The identification and support of both used *and* unused capacities of institutions, communities, households and individuals is crucial for sustainable risk reduction and adaptation planning.

The terms coping capacity and adaptive capacity are often used as synonyms for the concept of capacity just described and commonly refer to both used and unused capacities. Coping capacity is defined by UNISDR (2009: 8) as '[t]he ability of people, organizations and systems, using available skills and resources, to face and manage adverse conditions, emergencies or disasters'. Whilst the term adaptive capacity is not included in UNISDR's glossary (UNISDR 2009), a definition can be found in the introduction to the IPCC Fourth Assessment Report, stating that 'adaptive capacity is the ability of a system to adjust to climate change [including climatic extremes and variability] to moderate potential damages, to take advantage of opportunities, or to cope with the consequences' (IPCC 2007a: 21). On the basis of a comprehensive understanding of disaster and disaster risk, it can be argued that people's adaptive capacity and people's coping capacity are determined by the same attributes or factors. Both terms are used as synonyms in this volume.

Income level is often considered to be the most important factor shaping the adaptive capacity of individuals, households and communities because economic resources can make people more likely to succeed in safeguarding their lives, property and livelihoods[2] – even though their economic losses in disasters are often of greater magnitude in absolute numbers. A higher income may, for instance, allow people to live and pay rent in a safer area or to invest in costly risk-reducing and adaptation measures. Similarly, economic resources such as savings and insurance can enable a faster recovery in the aftermath of disasters.

New research, however, challenges the understanding of income (or economic resource availability in general) as the single or most important factor in adaptive capacity.[3] People's level of formal education is, for instance, increasingly seen as another or even a more important key factor (Box 2.1).

Further factors that shape individuals' adaptive capacity are, amongst others, gender (see Box 2.14), family structures, community cohesion and social networks, health status, age, existing traditional values and ideological beliefs, access to basic human rights, (in-)equality, levels of corruption, compliance with laws, as well as access to information, adequate housing, infrastructure, technology and institutional services. The case of the 2003 European heatwave cruelly illustrates how individuals' vulnerability can be shaped by factors like gender, community cohesion, social networks and age. Many victims were elderly people who lived alone in neighbourhoods that lacked a sense of community. They were trapped inside their homes and some were found dead only after several days (IPCC 2007a; Jha 2006; Larsen 2006; see also Section 1.1). The Chicago heatwave in 1995 is another of many examples, which shows that it is generally the elderly, poor and isolated who are most affected (Dorell 2012). Notably, while the large majority of the deceased in the Chicago heatwave were male, in Paris (severely struck by the European heatwave in 2003), the opposite was found, that is to

BOX 2.1

Key factors influencing people's adaptive capacity

In response to the urgent need for a better understanding of the factors that shape people's capacity to cope with and adapt to adverse climate conditions, a research project coordinated by the International Institute for Applied Systems Analysis (IIASA) set out to investigate what role people's level of formal education plays in this context. One of the project components (carried out during 2010–2012) focused on assessing how formal education influences the adaptive capacity of the residents of two low-income settlements where climate-related disasters are recurrent: Los Manantiales in San Salvador (El Salvador) and Rocinha in Rio de Janeiro (Brazil).

In both case study areas, it was found that the average levels of education were lower for households living at high risk as opposed to residents from lower-risk areas. The study found the influence of people's educational level on their adaptive capacity to be twofold due to (a) its direct effect on aspects that reduce risk and (b) its mitigating effect on aspects that increase risk. In fact, on the one hand, formal education was identified to have a positive effect on issues such as people's level of awareness and understanding of existing risks, their access to information on potential risk reduction measures, possibilities of attaining a formal job and their interest in moving out of a risk area. On the other hand, formal education was also identified as having the potential to reduce underlying risk factors such as poor health, organized crime, corruption, teenage pregnancy, single motherhood and informal settlement growth,[a] including the stigmatization of slum dwellers, exclusion from formal decision-making processes, insecure tenure and inadequate housing and infrastructure. The results further suggest that education plays a more determinant role for women than for men as regards their capacity to adapt. (See Wamsler *et al.* 2012; see Box 5.5.)

[a] For simplicity, the terms informal settlement, marginal settlement and slum are used as synonyms in this volume. A slum can be defined as a settlement which lacks a combination of the following conditions: easy access to safe and affordable water; access to adequate sanitation; structural quality/durability of dwellings; sufficient living space; and security of tenure (UNHABITAT 2006).

say, the larger number of heat-related deaths in Paris occurred among elderly women (Cadot *et al.* 2007). This indicates that gender-based sociological factors in each region may be just as important as any biological differences. Further analyses of the Paris case have shown that the higher death risk associated with being female did not apply for the Parisian immigrant

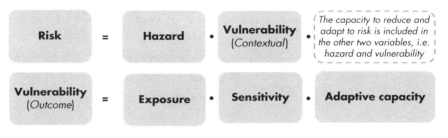

Figure 2.4 Making sense of seemingly contradictory concepts used in disaster risk- and climate change-related literature.

women, of whom many had roots in Africa and Asia. The reason for this is believed to be the fact that they more often live in multigenerational families (Cadot *et al.* 2007).

From a city perspective, the determinants of adaptive capacity are often summarized as being infrastructure, economic resources, technology, institutions and governance structures, equity, information and skills (Smit *et al.* 2001), which are influenced by individuals' adaptive capacity. Also in this context, formal education can be an important influencing factor because it has been related to most of the determinants of cities' adaptive capacity (e.g. to knowledge production, economic growth and as being a necessary condition for democratic institutions to develop and persist [Lutz 2010]).

Adaptive capacity is closely related to the concept of vulnerability described previously. The more adaptive capacity a person or a system has, the less vulnerable they are. Both factors are thus determined by virtually the same indicators. This is also illustrated in Figure 2.4 (upper part).

Before moving on to the term resilience, listed in Table 2.1 (first column), it is crucial to highlight that the terms vulnerability and capacity are differently presented in disaster risk- and climate change-related literature. Whilst this might be confusing, it is relatively easy to make sense of the seemingly contradictory concepts used.

It is crucial that urban actors are aware of the fact that the term vulnerability, when used in climate change-related literature, can be seen as a synonym for the concept of risk presented in Figure 2.4.[4] Climate change-related literature generally presents vulnerability as a function of exposure, sensitivity and adaptive capacity (e.g. Brenkert and Malone 2005; IPCC 2001, 2007a, 2012a; Morss *et al.* 2011; Smit and Wandel 2006). How can this be related to the risk equation (see Equation 1)?

In simple terms, the term exposure used in climate change-related literature refers to the existing hazards that a specific area is exposed to. Sensitivity is used to describe vulnerability patterns (including its three components, as described previously). As regards the term adaptive capacity, the only difference is that it is used as a separate variable in climate change-related literature, whilst in risk-related literature it is seen as an inherent component of the hazard and vulnerability concept (see Figure 2.4 and the description of the concept of adaptive capacity beginning on page 27).

Finally, in order to deal with the different connotations and use of the term vulnerability, in climate change-related literature sometimes a difference is made between outcome vulnerability and contextual vulnerability (Hamza et al. 2012; O'Brien et al. 2007). Outcome vulnerability is here understood as the result of projected impacts of climate change on a particular exposure unit (e.g. an urban community), which can be offset by adaptation measures. This understanding of (outcome) vulnerability is comparable to the risk concept used in this book and illustrated in Figure 2.4. In contrast, contextual vulnerability describes the context in which climate change takes place; that is, the present inability to cope with external stressors, which relate to past damages, existent needs, challenges and capacities of the exposure unit in question (Figure 2.4). Due to the fact that the terms risk and (outcome) vulnerability are in parts used as synonyms, related indices consequently use quite similar criteria.[5]

Now we come to the terms resilience and transformation, the last words listed in the first column of Table 2.1. Before continuing to read, ask yourself: how is resilience linked to the other concepts already presented?

The use of the term resilience has become increasingly prominent, not only in disaster risk-related literature. More and more articles and books are written about its meaning and its historical development (e.g. Béné et al. 2012; Cannon and Müller-Mahn 2010; Comfort et al. 2010; Zolli and Healy 2012). In general, resilience is an attribute of a system's behaviour, indicating how well the system performs. From a disaster risk perspective, resilience refers to 'the ability of a system, community or society exposed to hazards to resist, absorb, accommodate to and recover from the effects of a hazard in a timely and efficient manner' (UNISDR 2009: 24). In line with this, and in accordance with the concepts described previously, disaster resilience can be seen as the *antithesis* of risk: the more disaster resilient a place is, the less risky it is (see Boxes 1.4 and 2.2).[6] Consequently, risk is generally discussed in relation to particular population groups, sectors or places, whereas resilience is more discussed in relation to what helps to

BOX 2.2

Resilience – the roots of the concept and its emergence

The term resilience has a number of academic roots because it has been used in engineering (where it relates literally to the ability of a structure to 'resile' – to recover its original shape after being distorted by a stress), psychology (relating to a person's capacity to deal with stresses) and systems thinking (especially in relation to ecosystems). The root that is most related to the disaster and climate change context is derived from the last of these and is normally traced to the work of Holling in the 1970s (e.g. Holling 1973). The analogy with engineering is that an ecosystem has the ability to 'bounce back' to an arguably equilibrium state despite being disturbed by natural or human-caused stresses. By extension to human systems (often characterized as being embedded in nature as 'socio-ecological systems'), the idea of resilience is that they may be able to absorb disturbances and continue to function by realigning themselves to carry out basically the same functions. This aspect has received considerable attention in recent years in relation to climate change and the need for resilience to deal with it. Such is the appeal of the concept in this framing that major international and national institutions (including the World Bank, Britain's foreign aid ministry and the Department for International Development) have adopted it with enthusiasm.

However, the concept does have controversial aspects. It is largely replacing the terms 'risk' and 'vulnerability' in development, disaster and climate change discourses. Some argue that this is because it is politically more neutral and therefore useful to those who want to avoid discussing political and economic causes of human problems. It is difficult to be 'against' the idea of resilience and so it is convenient and sounds good. Building systems and people to be resilient surely does no harm. But it also detracts from understanding why people need to be resilient: it is economic and political systems that have caused climate change, with very specific and identifiable perpetrators of the problem. Ecosystem damage often has clear causes that are also linked to corporate or government actions that exploit resources and the environment. To understand this, it is necessary to conduct risk and vulnerability analyses, precisely to find what factors are leading some groups of people and/or environments to be harmed. In its current usage, resilience becomes a concept in which it is possible to avoid the need to address causes by focusing instead on the need to build capacities without having to explain why the problems exist.

By Terry Cannon

protect them. Importantly, however, this understanding does not imply that risk reduction and adaptation in cities are one-off actions that have as their only goal that of eliminating urban risk. They are, in fact, processes that work towards resilience and sustainability by expanding the focus from trying to prevent, control or resist disasters and hazards (including climatic extremes and variability) to a broader systems resilience framing in which we learn how to live and cope with an ever-changing, and sometimes risky, environment (Cannon and Müller-Mahn 2010; Morss *et al.* 2011).[7] This might require not only incremental but also transformative improvements, especially in a context of climate change. The magnitude of changes (such as greater frequency and severity of extreme events, trends in temperature and shifts in seasonality) means that an incremental approach may become no longer viable.

BOX 2.3

From resilience to transformation

Resilience has established itself in the language of risk reduction, climate change adaptation and as a host of other areas without any shared idea of its meaning or implications for sustainable development. It has become assumed as a normative good. The term resilience has many origins, but has often been interpreted in practice for urban risk reduction and adaptation as little more than a repackaging of 'return to normality' or as business-as-usual for engineering in communities which have long applied the term to indicate norms such as fail-safe design. This is really a missed opportunity. The concept of transformation is a response to the ongoing resilience debate (Pelling and Manuel-Navarrete 2011). 'Aired' through the SREX report (IPCC 2012a), it has begun to develop some momentum (see Kates *et al.* 2012; O'Brien 2011). Resilience is here positioned as a policy space for critical reflection and the consideration of uncertainty, one that can then flow into actions described as incremental (improving the efficiency of existing risk management approaches to maintain systems functions) or transformative (accepting and even provoking systems change for long-term sustainability). Transformation thus opens new possibilities for moving beyond planning for the status quo, to a questioning of development values and practices.

By Mark Pelling

With the growing awareness that systems might have to transform in order to be able to resist increasing disasters and climate change in the long term, the notion of transformation has become more and more prominent (Box 2.3). Transformation can be defined as 'the altering of fundamental attributes of a system (including value systems; regulatory, legislative, or bureaucratic regimes; financial institutions; and technological or biological systems)' (IPCC 2012b: 564). Whilst it might be questionable to add, again, another term to the conceptual jungle associated with risk reduction and adaptation (see Figure 2.1 and Table 2.1), it might be beneficial for city authorities and planners.

Applied to the urban setting, the emerging concept of 'sustainable urban transformation' has indeed resulted in increasing emphasis placed on understanding cities as a source of possibilities, promoting active collaboration among diverse stakeholders (including researchers and practitioners) and integrating different perspectives and bodies of knowledge and expertise (McCormick *et al.* 2013). As regards the latter, it has the potential to foster further rapprochement between risk reduction and adaptation professionals by (better) defining their aim and the processes in which they need to (jointly) engage (see Section 2.3). This is also because the notion of transformation links back to the idea of addressing the root causes of risk, which is advocated by risk reduction professionals and is 'a key premise of the Hyogo Framework for Action' (UNISDR 2012d; see Boxes 2.5 and 2.7). Root causes can here be defined as an interrelated set of structural factors and processes within a society, which often have arisen in another time (i.e. in past history) or another place (i.e. in a distant centre of economic or political power) and are typically so entrenched in today's society that they become 'invisible' and thus hard to detect.[8] Addressing root causes thus often implies the use of not only incremental, but also transformative improvements.

Based on the resultant understanding of the different concepts that underlie urban risk reduction and adaptation (listed in Table 2.1, first column), we can now move to more operational aspects that are of direct relevance for city authorities' and planners' daily work (see Sections 2.2–2.3). The concepts of risk reduction and adaptation and the associated features that make cities resilient are described in detail in the following sections.

2.2 Measures for reducing and adapting to increasing risk

Whilst a disaster is said to be the result of 'insufficient capacity or measures to reduce or cope with potential negative consequences' (UNISDR 2009: 9), the commonly used risk equation (see Equation 1) does not explicitly include such capacities or measures (see Figures 2.2 and 2.4). It is a simplified presentation of risk, which provides a basic conceptual framework that clearly illustrates that (a) there are no such things as 'natural' disasters in cities, and (b) urban planning, through its influence on hazards and vulnerability, is capable of counteracting disasters and climate change impacts, but also of strongly reinforcing them (see Boxes 1.1–1.2). For city authorities, aid organizations and planners, this framework is, however, of little operational value because it does not provide an adequate planning basis for assessing urban risk and identifying potential measures to reduce and adapt to the identified risk.

A more operational framework needs to link the different risk components directly to corresponding risk reduction and adaptation measures. In accordance with the risk factors presented in the previous section, this has to include measures to address (a) existing hazards and (b) existing vulnerabilities that make it difficult for an area to withstand, cope with and recover from hazard impacts. Such measures can be called:

1. Hazard reduction and avoidance (to reduce or avoid current and future hazard exposure)
2. Vulnerability reduction or disaster mitigation (to reduce current and future susceptibility in order to better withstand potential hazard impacts)
3. Preparedness for response (to create functioning and flexible mechanisms and structures to respond to potential hazards/disasters)
4. Preparedness for recovery (to create functioning and flexible mechanisms and structures to recover from potential hazards/disasters).[9]

If we express this more operational understanding of risk in the form of a pseudo-equation, risk (R) can be expressed as the product of hazard(s) (H), vulnerability (V) and lack of mechanisms for response and recovery (LR), which is divided by the corresponding risk reduction and adaptation measures: hazard reduction and avoidance (HR), vulnerability reduction (M) and preparedness for response and recovery (PP) ($R = H/P \cdot V/M \cdot LR/PP$). This illustrates the fact that an increase in hazard(s), vulnerability or lack of response and recovery results in increased levels of risk, whilst an increase in measures for hazard reduction or avoidance, vulnerability reduction and/or preparedness results in reduced levels of risk.

The four different measures listed can be initiated by (a) institutions or (b) single individuals, households or communities. As regards the latter, people's own way to reduce and adapt to risk is often referred to as local coping strategies, especially in disaster risk-related literature (see Table 2.1, second column).

BOX 2.4

Measures for risk reduction and adaptation – defined

There are four different measures to reduce and adapt to urban risk:[a]

1. **Hazard reduction and avoidance** aims (to increase the capacity) to reduce or avoid existing or likely future hazards that threaten communities, households and/or institutions.
   Note: The term hazard in this instance includes climate- and non-climate-related hazards, with climate-related hazards referring to both climatic extremes and variability. Further note that related measures are sometimes also called disaster prevention.

2. **Vulnerability reduction** or disaster mitigation aims (to increase the capacity) to minimize the existing or likely future susceptibility of communities, households and/or institutions in order to better withstand potential future hazard occurrence/disasters.
   Note: The term mitigation in this instance stands for disaster mitigation, not for climate change mitigation. Climate change mitigation aims at reducing greenhouse gas emissions to minimize climate change. In the context of disaster risk management, such activities can be categorized as disaster prevention.

3. **Preparedness for response** aims (to increase the capacity) to establish effective response mechanisms and structures for communities, households and/or institutions so that they can react effectively during and in the immediate aftermath of potential future hazards/disasters.
   Note: Preparedness for response includes, for instance, early warning, contingency and evacuation planning.

4. **Preparedness for recovery** aims (to increase the capacity) to ensure appropriate recovery mechanisms and structures for communities, households and/or institutions that are accessible after potential future hazards/disasters.
   Note: Preparedness for recovery is crucial for risk reduction and adaptation because (a) both spontaneous and planned early recovery starts the moment a hazard occurs, (b) risk areas affected by a hazard are generally still in the process of recovering from earlier hazards, and (c) the term hazard includes primary and secondary hazards and not only rapid but also slow-onset events which can develop over time or are successive (e.g. aftershocks) (Wamsler *et al.* 2012). Preparedness for recovery includes formal and informal ways of risk financing (i.e. risk

transfer and sharing through different types of disaster insurance mechanisms). At household level, the term 'self-insurance' (instead of preparedness for recovery) is also used and can be defined as 'the creation or maintenance of formal or informal security systems that help people access financing sources or mutual social help in the event of a disaster' (Wamsler 2007a: 122).

Risk assessment can be seen as an inherent part of all the measures listed in this box and as an inevitable precondition needed in order to identify and design adequate measures to reduce and adapt to risk.

---

[a] Note that in the mainstreaming context, using the term 'measures for risk reduction and adaptation' is somewhat misleading because risk reduction and/or adaptation are generally not the main or single objective of related measures. The wording is, however, retained in this book for the sake of simplicity (see Section 2.4).

Other terms with similar connotations are: individual practice for risk reduction, private adaptation, autonomous adaptation, adaptive response, adaptive behaviour and adaptive practice. For the coping strategies of rural societies, the term indigenous practice is also used.[10]

The definitions of the four risk reduction and adaptation measures are presented in Box 2.4. They highlight that, for each type of measure, urban actors always have two different options for assisting people to reduce risk and adapt to changing climate conditions. These are:

1. Directly – by reducing the corresponding risk component
2. Indirectly – by increasing capacities to reduce the corresponding risk component, thus enabling individuals, households, communities, institutions and ultimately societies to reduce their level of risk on their own.

Both options can be implemented either with or without the active participation of the different stakeholders involved. In addition, both options can be built on existing (used and unused) capacities or, alternatively, create new capacities or separate structures (see Section 2.1 on adaptive capacities). Nevertheless, it has to be noted that both the active participation of institutions and at-risk people *and* the building on their respective capacities has proved to be crucial for achieving sustainable change.

The four risk reduction and adaptation measures listed in Box 2.4 need to be based on continuous risk assessments, monitoring and learning mechanisms

(Becker *et al.* 2011). Risk assessment is thus not a risk reduction or adaptation measure (i.e. a measure that in itself leads to a reduction of risk), but an inevitable precondition needed to identify and design adequate measures to reduce and adapt to increasing risk (see Boxes 2.2 and 2.4).

City authorities and planners need a basic understanding of the risk assessment process in order to be able to critically review existing assessment reports, use their outcomes, update them if needed or assist in the proper elaboration of new assessments. So what is a risk assessment? Before continuing to read, ask yourself: how can risk assessments be conducted and what do they include?

In simple terms, risk assessment is the process of analysing and evaluating risk. It is composed of a risk analysis followed by a risk evaluation. The risk analysis aims at determining the likelihood that a specified negative event will occur (ISO 2009). It thus has to provide answers to the following three questions:

1.   What can happen?
2.   How likely is it to happen?
3.   If it happens, what are the consequences?

In order to be able to answer these three questions for a specific urban area, the following aspects need to be identified:

*   The issues that are valuable and important to protect in the specific location (e.g. livelihoods or health status of population, specific population groups and specific services or infrastructure)
*   Specific hazard events that can have an impact on the aspects to be protected
*   The degree of vulnerability of these aspects to the impact of each hazard event (including location-specific conditions and existing capacities to respond to and recover from the event).

The risk analysis is followed by the risk evaluation, which is the process of evaluating the results of the risk analysis in order to determine whether the identified risk is tolerable or not (ISO 2009). If the identified risk is not tolerable, the level of tolerable risk has to be defined. On this basis, adequate risk reduction and adaptation measures have to be designed and implemented with the aim of reducing the identified risk to the tolerable risk level. It is crucial to reach a consensus on the level of risk that is acceptable. If possible, related criteria should be determined in consultation with subject matter

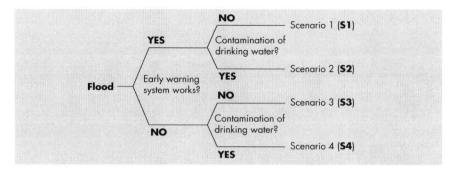

What can happen? (Scenario)	How likely is it? (Estimated frequency)	What are the consequences? (Estimated number of fatalities)
S1: Flood where the early warning system works and there is no contamination of drinking water.	**Very likely** (More than once in every 10 years)	**Minimal** impact (No fatalities)
S2: Flood where the early warning system works but drinking water is contaminated anyway.	**Likely** (Once per 10–100 years)	**Major** impact (6–20 fatalities)
S3: Flood where the early warning system does not work; nevertheless, there is no contamination of drinking water.	**Very likely** (More than once in every 10 years)	**Minor** impact (1–5 fatalities)
S4: Flood where the early warning system does not work and the drinking water is also contaminated.	**Unlikely** (Once per 100–1,000 years)	**Catastrophic** impact (>100 fatalities)

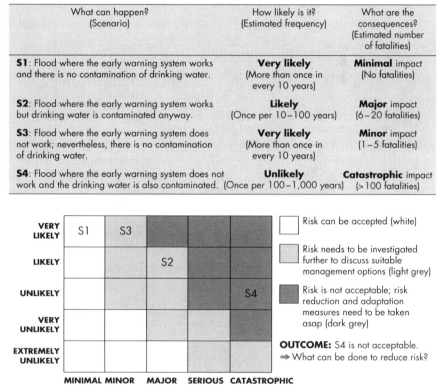

Figure 2.5 An example of a simple scenario tree and related risk matrix.
Source: adapted from Tehler (2013).

experts as well as the general public. This was, for instance, achieved in North Vancouver, Canada where a task force of volunteer residents was formed to recommend an acceptable level of risk in their communities. The outcomes have then been incorporated into the city's official community plan, strategic planning and development permit processes (UNISDR 2012b).

On the basis of the described risk assessment process, the following aspects should be part of every basic risk assessment:

1. Description of the identified values (i.e. the issues that are valuable and important to protect and which guide the focus of the risk assessment)
2. Simple map (or structure model) of the risk area to illustrate its key elements and their linkages (e.g. flood levels, key infrastructure, identified values)
3. Hazard analysis
4. Vulnerability and capacity analysis
5. Risk scenarios, which show what can happen, how likely this is and what the consequences will be. Scenario trees, also known as event trees, can be used for structuring the risk scenarios (Figure 2.5)
6. Summary of results, which links the different analyses with the identified values. This can be done with the help of risk matrices where each risk scenario can be represented as a combination of likelihood and consequence, also indicating which risk scenarios are acceptable and which are not (Figure 2.5).[11] Other ways to present the results are the use of FN-curves and risk contours (see Tehler 2013).

Tables 2.4 and 2.5 were designed to provide a basic framework that can assist urban actors to systematically analyse existing hazards, vulnerabilities and

Table 2.4 Basic framework for systematic hazard, vulnerability and capacity analysis in urban areas (some examples are scribbled in the table to illustrate its use)

Hazards	Vulnerabilities[a] (of city dwellers and the institutions serving them)	Used capacities[a] (existing risk reduction measures – including local coping strategies)	Correlation between hazards, vulnerabilities and (lacking) capacities
Floods Land-slides	Precarious houses No land ownership Dependency on one main road Many elderly people Persistent damages from past disasters Low levels of literacy	Functional emergency committees Functional early warning system Backup generators for emergency power Local initiatives to stabilize slopes Community house offers a refuge	Precarious houses are built by informal builders without professional training Use of plants which do not support the stabilization of slopes Few measures/capacities for recovery

Note:
[a] Aspects to be considered: physical, environmental, social, economic, political and institutional. Further note that vulnerabilities and related capacities are different before, during and after hazard/disaster impacts, which needs to be considered here.

Table 2.5 Framework for systematic vulnerability and capacity analysis to identify adequate risk reduction and adaptation measures (some examples are scribbled in the table to illustrate its use)

Relevance of capacity for	Used capacities		Unused capacities	
	Citizens	Institutions	Citizens	Institutions
Hazard reduction and avoidance[a]	Planting of slopes (but with inadequate plants) Landslide retaining walls built of old tyres (well maintained and tyres available)		Good community organization/ local committee interested in working on hazard reduction	Reforestation and water management project in neighbour community that could be expanded to this area
Vulnerability reduction[a] (conditions of location pre-disaster)	Community house also promotes social cohesion	Main road with adequate drainage and related maintenance	Informal builders who could be trained in disaster-proof construction Possibility for households to access legal tenure if practical issues are supported	NGOs with capacities to also provide courses on vulnerability reduction
Preparedness for response[a]	Local emergency committees that help organize evacuations, etc.	Community house used as shelter during evacuations Backup generator to provide emergency power	Majority has TV, radio and mobile phones: expandable channels for warnings, campaigns, etc.	Church bells could be used for disaster alarm

(continued)

Table 2.5 (continued)

Relevance of capacity for	Used capacities		Unused capacities	
	Citizens	Institutions	Citizens	Institutions
Preparedness for recovery[a]	8% of households have formal insurance	Advisory services for reconstruction	Social cohesion which could be fostered for sharing of resources, work, etc. during reconstruction	Existing microfinancing organizations that could engage in disaster/homeowner insurance Advisory services for reconstruction offered are relatively unknown within the community
General[a]	Many elderly people with knowledge on traditional coping strategies		Rising educational levels among younger generation (but still many avoidable drop-outs)	

Note:

[a] Aspects to be considered: physical, environmental, social, economic, political and institutional. Further note that capacities are different before, during and after hazard/disaster impacts, which needs to be considered here.

capacities, which are the central analyses of the risk assessment process (see third and fourth points listed previously). Table 2.4 can be used for the initial risk assessment. In the first column are listed the hazards to which the area or households in question (i.e. the valuable issue to be protected) are exposed, for example floods and/or landslides. Under the next heading, related vulnerabilities are specified. These can, for instance, include issues such as insecure constructions; no land ownership; loose high voltage transmission cables; power systems vulnerable to floods; potential pollution due to lack of sewage and waste management; low levels of literacy and income; or dependency on one main road with no secondary roads (to commute to larger cities and for emergency access). Subsequently, currently used capacities are included, which can be aspects such as the existence of functioning emergency committees, an alarm for early warning or backup generators to provide power in case of grid failure. Once all the aspects related to hazards, vulnerabilities and used capacities have been listed in the first three columns, it is helpful to re-check whether non-physical aspects have been analysed sufficiently. Once the first three columns are adequately filled in, the identified used capacities/ measures can be further analysed to assess the correlation between the identified hazards, vulnerabilities and capacities (e.g. it might be possible to identify that existing measures are dealing with only some specific risk factors or focus on physical aspects only) (Table 2.4, last column).

On the basis of this initial risk assessment, Table 2.5 allows more in-depth analyses of the existing capacities, including both used and unused capacities and their relation to the four potential measures to reduce and adapt to risk. Such in-depth analysis is crucial for urban actors to allow them to identify potential risk reduction and adaptation measures, which can potentially build on existing capacities. Table 2.5 was also designed because of the fact that there are hardly any tools for risk assessment that explicitly assess unused capacities. An exception is the so-called vulnerability and capacity analysis (VCA) promoted by the Red Cross (see IFRC 2007, especially Annexes 1 and 2: Tables 1.1, 1.2, 2.1 and 2.2). Unused capacities can be, for instance, the existence of many (but not formally trained) builders, or existing microfinancing institutions that could (but do not yet) offer disaster insurance. General issues that can positively influence people's adaptive capacities, such as levels of income, education or health status, can be listed in the last row (Table 2.5). Note that it is important to always include and interrelate the listed physical, environmental, social, economic and political/institutional aspects in the analyses. As an example, unused financial capacities might, for instance, be supported in a way that will permit physical vulnerabilities to be tackled (see Table 2.5).

After the risk assessment, the process there begins of identifying and designing adequate measures for risk reduction and adaptation in order to reduce the identified risk to the tolerable risk level. For the comprehensive identification and design of adequate measures, it might be helpful to systematically go through all physical, environmental, social, economic and political/institutional issues relevant to urban risk reduction and adaptation (thereby also considering urban–rural linkages). These issues include housing, land tenure, recreation and cultural heritage, transportation infrastructure, water and sanitation, energy systems, telecommunication, environment and natural resource management, waste management, biodiversity, social and public services (including education, healthcare, police and emergency services), economic sectors (such as industry, agriculture, national and municipal economy, livelihoods and informal sectors), rule of law, democracy, and institutional capacity and operational management.

During the process of identifying and designing adequate measures, the following criteria can be used to select and prioritize among different options:

1. Expected (long-term) impacts of action:
   - Reduction in number of deaths, injuries, property and economic losses
   - Reduction of other impacts on people's livelihoods, assets, etc.

2. Probability that the action will be implemented:
   - Social acceptance
   - Technical feasibility and maintenance
   - Administrative issues (e.g. organizational structures, staffing)
   - Political and public support
   - Legal viability
   - Economic issues (e.g. cost-effectiveness, available funding)
   - Environmental issues (e.g. long-term versus short-term benefits).[12]

Measures that build on existing (used and unused) capacities are most likely to fulfil most of the listed criteria. The analysis of these capacities is thus crucial.

2.3 Processes and working fields related to risk reduction and adaptation

After having discussed the underlying concepts, activities and measures of urban risk reduction and adaptation (see Sections 2.1–2.2),[13] we can now start linking them to the related processes and working fields (Table 2.1, third column).

As described at the beginning of this chapter, disaster risk management includes three processes, namely (a) disaster response, (b) disaster recovery and (c) disaster risk reduction. The third process, disaster risk reduction, is closely interlinked with climate change adaptation. Before continuing to read, ask yourself: what makes risk reduction and adaptation so different from the other processes listed in Table 2.1? How are they interlinked and how do they differ from each other?

In comparison with the longer tradition of disaster risk reduction, climate change adaptation is a relatively new lens through which to understand, assess and tackle risk and vulnerabilities. Box 2.5 describes the historical development of risk reduction and adaptation, showing that, until recently, they have in the main developed independently. Due to these historical developments, the understanding of different concepts such as risk reduction has changed considerably over the past decade. An awareness of these changes is needed by urban actors when reading literature from different (publishing) years.[14]

Today, the need for greater integration of risk reduction and adaptation (as promoted in this book) is widely recognized and supported, and their key commonalities and overlaps are receiving increased attention. Table 2.6 lists the most important similarities and overlaps, as well as existing differences. The listed differences can translate into a series of challenges when bringing together the previously rather isolated risk reduction and adaptation professionals, as highlighted in Box 2.6.

Importantly, risk reduction and adaptation share the aim of reducing the frequency and impacts of climate-related disasters and associated risk, which includes climatic extremes and variability (Table 2.6; see Boxes 1.2 and 1.3). In fact, risk reduction and adaptation to climate change cannot be meaningfully conducted without considering climate variability (Klein and MacIver 1999; IPCC 1997). In view of their shared aim, risk reduction and adaptation require virtually the same type of measures/activities (Table 2.6). These five main activities are (1) hazard reduction and avoidance; (2) vulnerability reduction; (3) preparedness for response; (4) preparedness for recovery; and (5) risk assessment (see Table 2.1, second column, bold terms and Box 2.4). In accordance with this understanding, UNISDR defines risk reduction as

> the concept and practice of reducing disaster risks through systematic efforts to analyse and manage the causal factors of disasters, including through reduced exposure to hazards, lessened vulnerability of people and property, wise management of land and the environment, and improved preparedness for adverse events.
>
> (UNISDR 2009: 10–11)

BOX 2.5

Historical development of risk reduction and adaptation: from stand-alone to cross-cutting

When disaster risk reduction (initially called disaster risk management) became recognized as a proper field of activity, it was understood as a new field of work. The first programmes on disaster risk reduction were thus implemented independently of existing response, recovery and development efforts, and their focus was mainly on improving societies' preparedness for response. In the late 1990s, such pilot programmes for disaster risk reduction emerged all over the developing world, for instance in Central America after Hurricane Mitch in 1998.

The lessons learnt from the pilot programmes led to a change in the original understanding of disaster risk reduction. While it was initially understood as a separate field of work, it was increasingly seen as a so-called cross-cutting or mainstreaming topic. First, interest and efforts moved from creating pilot programmes for disaster risk reduction towards integrating this new topic into disaster response. Shortly after, consensus emerged that it also needs to be integrated into recovery. The recovery phase was, in fact, identified as a better 'window of opportunity' for managing risk due to its longer time frames, closer cooperation with local stakeholders and linkages to long-term development. However, the focus on recovery did not last long (around 2001–2003).

In 2005, with the World Conference on Disaster Reduction held in Kobe (Hyogo Prefecture, Japan), the focus turned towards integrating disaster risk reduction into development. Interest had moved away from the focus on stand-alone programmes for disaster risk reduction to more challenging efforts of reviewing every single type of sectoral development work in order to make sure that risk is not increased and if possible reduced. The World Conference resulted in the so-called Hyogo Framework for Action (see Box 2.7), which is a confirmation of the described historical evolution of disaster risk reduction. The framework's first strategic goal states the need to integrate disaster risk reduction into development. Only the third goal relates to its integration into response and recovery.

When it comes to the relation between risk reduction and the more recent field of climate change adaptation, both fields have in the main developed independently. They are linked to two different professional expert communities, which, according to the policy documents they relate to, can be dubbed the Hyogo community and the Kyoto community (making reference to the Hyogo Framework from 2005 and the Kyoto Protocol from 1997). During the past few years, however, the two communities have been gradually moving closer. An illustration of this is the 'adaptation groups' of the IPCC (http://www.ipcc.ch), e.g. the group working on the *special report on managing the risks of extreme events and disasters to advance climate change adaptation* (*SREX*) (IPCC 2012a), which are made up of quite equal numbers of risk reduction and adaptation experts (see Box 2.6).

BOX 2.6

Working in the fields of disaster risk reduction *and* climate change adaptation: personal reflections

Having worked on issues related to disaster risk reduction for 40 continuous years, I had to make a radical 'change of gear' when I was asked by the IPCC to become a lead author of an IPCC report (IPCC 2012a). This project was the first 'official' attempt to bring together the previously rather isolated risk reduction and adaptation communities. For all the 246 scientists and practitioners involved in this mega-project, this became a steep learning curve and these are some personal reflections on the experience:

- The risk reduction community can draw from a long history while the climate change community has a comparatively short history of just about 25 years (climate change adaptation much less). This affects overall perspectives.
- Terminology presented us with many problems, for example the term 'mitigation' is used differently by both groups.
- Having previously worked within the under-resourced and often politically ignored world of disaster risk reduction, it was something of a revelation to become aware of the heavily politicized and far better funded world of climate change.
- Within the field of extreme climate events, risks are reasonably certain, being generally well established and calibrated, but in the arena of climate change major uncertainties still exist concerning scales of events and the nature, locations and impacts of climate change.
- Following the political fall-out after the press interception of emails in the Climate Change Centre of the University of East Anglia, the IPCC staff have become highly sensitive to further press 'invasions' or 'revelations' to a degree that has never been a concern within the disaster risk reduction field.

By Ian Davis

Table 2.6 Differences, similarities and overlaps between risk reduction and climate change adaptation[a]

Overlap	Differences
Shared aim: To reduce the occurrence and impacts of climate-related disasters and associated risk (including climatic extremes and variability). In this context, both RR and CCA focus on taking proactive steps towards reducing risk and adapting, instead of simply responding to events (see Box 1.3).	**Aim and activities:** CCA deals with climate-related events (i.e. climatic extremes and variability), whilst RR also covers non-climate-related events (such as earthquakes and volcanic eruptions) (see Tables 2.2–2.3).
Same activities: Risk reducing activities addressing climatic extremes and variability related to warmer- or colder-than-normal temperatures, droughts, windstorms, precipitation, floods and related landslides (see Section 2.2).	**Establishment of field:** RR and CCA are relatively new fields, with CCA being the more recent one. Consequently, in the field of RR there are more concrete experiences and lessons learnt (see Boxes 2.5 and 2.6).
Shared role: Mainstreaming issues to be integrated into response, recovery and long-term development (see Section 2.4).	**Historical development:** Whilst RR first focused on its integration into response, then recovery and now development, CCA has from the very beginning put its emphasis on the development context. There is thus little experience with CCA in the context of response and recovery (see Box 2.5).
Common stakeholders: Since around 2009, increasing overlap of professional groups dealing with RR and CCA.	**Time horizon:** CCA is based more on a future perspective and thus the analysis of potential future events. RR also considers future risk but is strongly based on a historical perspective and the analysis of past events. In addition, planning for CCA is often carried out with a longer time horizon.
Practice: At local household or community level, CCA measures and strategies are similar to, if not the same as, RR.	**Working approach:** RR is generally more event-focused and more based on bottom-up approaches than CCA.
	Political interest and financing: RR is considered to be less politically 'sexy' than CCA. Heads of State attend international CCA conferences and the level of funding is high, while support and funding for RR is average to low except after major disasters (see Box 2.6).

Table 2.6 (continued)

Overlap	Differences
	Professional profile: Probably more geographers, biologists, etc. in CCA and more disaster managers, engineers, architects and geophysical specialists in RR.
	Terminology: Many concepts (such as risk, vulnerability and mitigation) are used in both fields but with different connotations (see Chapter 2 introduction; Figure 2.4).
	Key organization: Key organization for RR is UNISDR; for CCA it is the IPCC and the United Nations Framework Convention on Climate Change (UNFCC) secretariat.
	Key policy documents: The Hyogo Framework from 2005 for RR and the Kyoto Protocol for CCA (see Section 1.1 and Box 2.7).
	Level of confidence and uncertainty: Uncertainty regarding future risk is highly discussed in the context of CCA, less in RR. This is because the existence of CC is still debated and even denied by some, whilst RR is not questioned and is rather met with indifference or apathy (see Box 2.6).
	Positive change: CCA focuses more explicitly on exploiting beneficial opportunities (e.g. longer growing seasons, increased heat for tourism, etc.) and thus includes not only communities that are negatively affected by climate change, but also those that may benefit from it.

Note:
^a RR: risk reduction, CC: climate change, CCA: climate change adaptation.

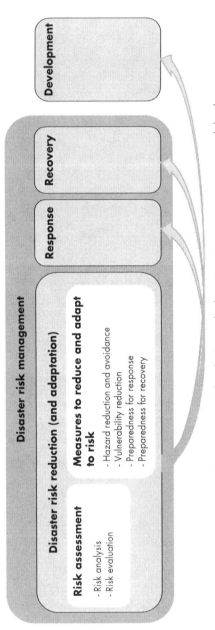

Disaster risk management

Disaster risk reduction (and adaptation)

Risk assessment

- Risk analysis
- Risk evaluation

Measures to reduce and adapt to risk

- Hazard reduction and avoidance
- Vulnerability reduction
- Preparedness for response
- Preparedness for recovery

Response

Recovery

Development

Mainstreaming risk reduction and adaptation into response, recovery and development

Figure 2.6 Activities which form part of risk reduction and adaptation and their mainstreaming.

In contrast, adaptation refers to 'adjustments in response to actual or expected climate and its effects' (IPCC 2012b: 556) and thus excludes non-climatic hazards and associated risk (Table 2.6).

The concepts of risk reduction and adaptation are very different from other processes, such as response and recovery, in the sense that they are so-called cross-cutting or mainstreaming issues that are of high importance during the whole disaster cycle (Table 2.6; see Section 2.4). What does this mean? It means that risk reduction and adaptation need to be integrated into urban planning and in all kinds of sector work before, during and after disaster occurrence:

1. In the pre-disaster context, i.e. in the context of development work and related everyday planning practice
2. During and in the immediate aftermath of disasters, i.e. in the context of disaster response
3. In the post-disaster context, i.e. in the context of disaster recovery.

A functioning risk management system and, ultimately, more resilient cities can thus only be achieved if risk reduction and adaptation are integrated into all three phases of development, response and recovery (Figure 2.6). This understanding is also supported by the strategic goals of the United Nations-endorsed Hyogo Framework for Action (Box 2.7), whilst most emphasis is today given to the pre-disaster or development context (Box 2.7, first strategic goal). The reason for focusing on the integration of risk reduction into development is simple: if, in the ideal case, urban planning and every kind of sector development work incorporated risk reduction and adaptation, disaster occurrence would be minimal and thus disaster response and recovery hardly necessary.

With both risk reduction and adaptation being mainstreaming issues, city authorities and planners need a basic understanding of (the key aspects, dif-ferences and interconnections of) the three potential mainstreaming con-texts if they want to adequately assist people at risk and promote sustainable change (see Figure 2.6). Before continuing to read, ask yourself: what is the difference between disaster response, disaster recovery and development work? And how do they relate to the other processes listed in Table 2.1 (namely humanitarian assistance, emergency management, disaster relief and rescue, reconstruction, rehabilitation, early recovery, technical assis-tance and technical cooperation)?

In every disaster-prone area, there is a constant succession and overlap of disaster response, recovery and development work, which is related to the

BOX 2.7

UNISDR and the Hyogo Framework for Action

Created in December 1999, the United Nations Office for Disaster Risk Reduction (UNISDR) is the secretariat of the International Strategy for Disaster Reduction (ISDR). It is the successor to the secretariat of the International Decade for Natural Disaster Reduction (1990–1999). UNISDR is the focal point in the United Nations system for the coordination of disaster reduction and adaptation and provides a vehicle for cooperation among governments, organizations and civil society actors to assist in the implementation of the Hyogo Framework for Action 2005–2015: Building the Resilience of Nations and Communities to Disasters (HFA).

The HFA is a 10-year plan to make the world safer from natural hazards. Its aim is a 'substantial reduction of disaster losses, in lives and the social, economic and environmental assets of communities and countries'. It was adopted by 168 Member States of the United Nations in 2005 at the World Disaster Reduction Conference. Its three strategic goals are:

1. The more effective integration of disaster risk considerations into sustainable development policies, planning and programming at all levels, with a special emphasis on disaster prevention, mitigation, preparedness and vulnerability reduction;
2. The development and strengthening of institutions, mechanisms and capacities at all levels, in particular at the community level, that can systematically contribute to building resilience to hazards;
3. The systematic incorporation of risk reduction approaches into the design and implementation of emergency preparedness, response and recovery programmes in the reconstruction of affected communities.

The HFA expires in 2015. The post-2015 framework explains what is to come.

By Helena Molin Valdes. See also http://www.unisdr.org

so-called disaster cycle.[15] In simplistic terms, response is implemented during and in the immediate aftermath of a disaster to save lives and meet basic subsistence needs. In contrast, the recovery phase has the objective of restoring and also improving people's former living conditions. It further provides the foundation for the (re-)introduction of development work, which aims at increasing people's living conditions and quality of life in the longer-term, for

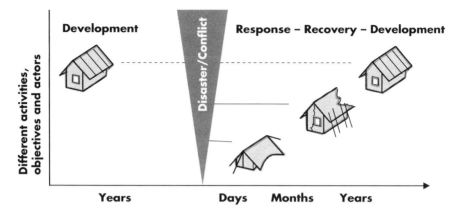

Figure 2.7 Simplistic sequence of disaster response, recovery and development.
Source: adapted from Wamsler (2002).

instance by sustainably reducing urban segregation and poverty (Figure 2.7). Although there are strong links and overlaps between all three phases, several characteristics, such as the specific objectives, activities and the actors involved, generally differ to a considerable extent and are highly context-specific.

More precisely, disaster response can be defined as 'the provision of emergency services and public assistance during or immediately after a disaster in order to save lives, reduce health impacts, ensure public safety and meet the basic subsistence needs of the people affected' (UNISDR 2009: 24). Disaster response is thus predominantly focused on immediate and short-term needs and includes both life-sustaining measures (i.e. disaster relief) and life-saving measures (i.e. rescue). In practice, response measures, such as the supply of temporary housing and water, may, however, extend far into the recovery and development stages. Disaster response can also be called emergency management or humanitarian assistance (see Table 2.1). Consequently, the terms 'humanitarian' and 'humanitarians' always make reference to the concept of humanitarian assistance and, thus, the context of disaster response. Measures taken to prepare for disaster response are called preparedness for response and can include activities such as early warning, contingency and emergency planning (see Box 2.4).

In comparison with disaster response, disaster recovery is defined as the 'restoration, and improvement where appropriate, of facilities, livelihoods and living conditions of disaster-affected communities' (UNISDR 2009: 23). Because the recovery stage offers better conditions for addressing underlying risk patterns than the response context, it is often called the 'window of opportunity' for integrating risk reduction (see Box 2.5). This understanding is also in line with the 'building back better' principle, which is widely promoted in any recovery efforts (UNISDR 2009). Disaster recovery includes early recovery, rehabilitation and reconstruction (see Table 2.1). Rehabilitation generally refers to the restoration of former conditions, whilst reconstruction also includes the improvement of what has been destroyed or lost. Another common differentiation between the concepts of rehabilitation and reconstruction is when they are used to distinguish between constructive measures (of reconstruction) and non-constructive measures (of rehabilitation) to both re-establish and improve former conditions. Early recovery is a relatively new concept that promotes the need for initiating recovery efforts as soon as possible after disaster occurrence and to better link the processes of response, recovery and development (UNDP/CWGER 2008). Measures taken to prepare for disaster recovery are called preparedness for recovery and can include activities such as risk transfer and sharing (see Box 2.4 and Table 2.1).

Unlike disaster response and disaster recovery, development work does not necessarily have a direct link to hazard or disaster occurrence. Development work can be defined as long-term support that seeks to alleviate poverty, improve the living standard of the population, strengthen the economy and build capacities for good governance. Related activities are carried out in areas that are prone to disasters, as well as in areas that are not prone to disasters. In disaster-prone areas, development work should also address the underlying risk factors (of inadequate development), which led to and/or may lead to disasters or conflicts in the first place. Development work and related planning is generally carried out by local actors such as city authorities and local non-governmental organizations (NGOs). If local actors receive financial or technical support from other countries, terms such as technical assistance or development cooperation are used. Since the term cooperation implies a two-way relation of equal partners (as opposed to assistance, which implies a one-way relation of giving and taking), development cooperation is today the more dominant term.

Attention must be drawn to the fact that even those urban actors who only focus on development work require a basic understanding of the key aspects, differences and interconnections between disaster response, recovery and development because the mainstreaming of risk reduction and adaptation into development work also means the integration of measures to prepare for response and recovery (see Section 2.2 and Figure 2.6). Within the different contexts (i.e. response, recovery and development), risk reduction can (and has to) be mainstreamed in all kinds of sector work. This includes, for instance, housing, recreation and cultural heritage, transportation infrastructure, water and sanitation, energy systems, telecommunication, environment and natural resource management, waste management, economic sectors (such as agriculture, industry and informal sectors) and social and public services (including education, healthcare, police and emergency services). Section 2.4 will discuss the concept of mainstreaming in further detail.

Finally, and before moving on to the next section, it has to be noted that the possible combinations of the concepts presented in this and Section 2.2 are virtually endless. They are an expression of the growing interest in linking different fields, disciplines, processes and various actors' efforts to address increasing disasters and climate change. Illustrative examples are the emerging terms of development-oriented adaptation, development-oriented response, adaptive risk reduction, adaptive resilience, adaptive transformation, transformative adaptation, sustainable adaptation, sustainable risk reduction, resilient risk management, resilient adaptation and climate-smart disaster risk reduction or management. With the same idea, more and more related aspects, perspectives and 'lenses' to look through are added, such as social protection[16] and human security,[17] also resulting in new terms, such as adaptive social protection. Especially in urban contexts, there has lately been considerable research and associated extensive literature on disasters, climate change, and human security implications and interactions (e.g. Pelling and Wisner 2009; Redclift *et al.* 2011; Simon and Leck 2010).

2.4 Mainstreaming risk reduction and adaptation into urban planning

With risk reduction and adaptation being cross-cutting issues, city authorities and other urban actors not only have to be familiar with the different ways in which risk (associated with increasing disasters and climate change) can be reduced in a specific location. They also need to know how risk reduction and adaptation can be mainstreamed to become an integral part of urban planning practice in general (see Table 2.1, second column, last term listed). But what does this actually mean? Before continuing to read, ask yourself: what does mainstreaming mean? What comprises the concept of mainstreaming risk reduction and adaptation, and how can it be achieved? And is there a conceptual difference between mainstreaming and integrating?

The term mainstreaming generally signifies the modification of a specific type of core work in order to take a new aspect or topic into account and to act indirectly upon it.[18] Consequently, mainstreaming does not mean to completely change an organization's core work and responsibilities, but instead to view them from a different perspective and carry out any necessary alterations, as appropriate. It is thus not about 're-inventing the wheel', but about looking into what already exists so as to build as much as possible on existing structures, mechanisms and procedures. It is about 'picking people up from their doorsteps' (Jürgens 2011).

In contrast, the term integrating can be used in a broader sense, which means that mainstreaming can be understood as a specific way of integrating a new aspect or topic into a specific type of core work. The difference between mainstreaming and integrating can best be illustrated with some examples: imagine an organization whose core work is the construction of social housing for urban poor communities. In order to integrate a new aspect into the organization's work, let's say HIV/AIDS, the organization could start with distributing condoms to their project beneficiaries. If the new aspect to be integrated is not HIV/AIDS but risk reduction, the organization could decide to establish early warning systems in their project areas. In both cases, a new aspect was integrated into the organization's work by engaging the organization in a new field of activity, which is not part of its normal day-to-day operations. New activities have been added with the objective of directly acting upon the issue of HIV/AIDS or risk reduction, accordingly. The organization's core work itself has not been altered. Conceptually speaking, the organization has thus not engaged in the mainstreaming of a new aspect.

Whilst a differentiation between integrating and mainstreaming is not always clear-cut or easy to make, it is crucial for urban actors to understand the difference between revising and improving their day-to-day work (to take a new aspect into account) as compared with adding on new activities (which are not necessarily related to their core work).

Box 2.8 shows the two different strategies that can be adopted by city authorities, aid organizations and planners to integrate risk reduction and adaptation into urban development[19] at the local programme level. They can be implemented separately or in combination. The second strategy falls under the concept of mainstreaming and is the basic strategy of responding indirectly, that is, through an organization's sector-specific programme work. This strategy can be called programmatic mainstreaming (see Box 2.8, Strategy II). But what knowledge is needed to be able to implement this strategy? Or more specifically, what knowledge do city authorities and other urban actors need to modify their work in such a way as to reduce the likelihood of any programme measure increasing risk and, if possible, to maximize their potential to reduce risk?

In order to engage in programmatic mainstreaming (see Box 2.8, Strategy II and Table 2.7), urban actors need to be familiar with:

- The concept of mainstreaming
- The different ways in which risk can be reduced (see Box 2.4 and Figure 2.6)[20]
- The ways in which their work in urban development is interlinked with disasters and climate change (described in Chapter 3).

If city authorities and planners fail to consider programmatic mainstreaming, the outcome can be harmful or, at best, represent a missed opportunity to contribute to risk reduction and adaptation in a certain area. Importantly, engaging in programmatic mainstreaming does not always mean an increase in costs. In some instances it might, in fact, even result in doing less rather than more (for instance, fewer river regulations to reduce the risk of floods).

But is it sufficient to make risk reduction and adaptation an inherent part of some selected programmes? The ultimate goal of mainstreaming is to

---

BOX 2.8

From integration to mainstreaming: strategies with focus on the local programme level (e.g. urban development programmes)

Strategy I (also called add-on risk reduction or adaptation): implementation of specific programmes or programme components that are explicitly and directly aimed at reducing risk and are not part of the implementing body's sector-specific work.

Strategy II (also called programmatic mainstreaming): modification of sector-specific programme work in such a way as to reduce the likelihood of any programme measure actually increasing risk and to maximize the programme's potential to reduce risk.

BOX 2.9

Mainstreaming strategies with focus on the organizational functioning of implementing bodies (e.g. city authorities)

Strategy III (also called organizational mainstreaming): modification of the organizational management, policy, corpus of legislation, working structures and tools for project implementation – to ensure the integration of risk reduction and adaptation at the local programme level and to further institutionalize it.

Strategy IV (also called internal mainstreaming): modification of an organization's way of operating and of its internal policies so that it can reduce its own risk in terms of impacts created by disasters and climate change and ensure its continuous functioning.

achieve a culture of resilience, which can surely not be achieved simply by those means. Before continuing to read, ask yourself: how can city authorities and planners achieve a more sustainable mainstreaming of risk reduction and adaptation into all kinds of urban development work?

As a first step, city authorities and other types of implementing organizations need to establish an institutional setting which is able to support and back up all their efforts to integrate risk reduction and adaptation at the local programme level (i.e. Box 2.8, Strategies I and/or II). There are two different mainstreaming strategies to guide related changes at the institutional level (Box 2.9). Strategy III is titled organizational mainstreaming and aims at making the integration of risk reduction and adaptation a standard procedure of an organization's work. Related measures can include modifications of the organization's management, policies, corpus of legislation, working structures or tools for project implementation. Strategy IV, titled internal mainstreaming, has the objective of ensuring the continuous functioning of the organization, also in times of increasing disasters and climate change. In this context, implementing organizations need to consider the following aspects:

• Likelihood of damage to the building that houses the institution
• Compliance of building (institutional premises) with local construction standards

- Potential damage to office equipment (documents, computers, etc.), vehicles or other means of transport
- Risk of theft due to potential loss of keys or other security equipment
- Probability of looting if there is a breakdown in law and order
- Potential impacts on communication between staff
- Location of staff in high-risk areas
- Potential impacts on access to the organization's clients.

To achieve a culture of resilience, it is, however, not enough if only one or some of the organizations that serve urban populations have risk reduction and adaptation included as an inherent part of their work. There are two further mainstreaming strategies, which city authorities and other urban institutions can adopt, to positively influence urban sector work and policy overall (Box 2.10, Strategies V and VI). They are called interorganizational mainstreaming and educational mainstreaming. Both strategies relate to the cooperation with, and capacity development of, other urban stakeholders, which includes citizens, all kinds of implementing organizations and educational bodies. They comprise the promotion of more distributed risk governance systems, as well as better science–policy integration.

From the standpoint of a specific local organization, such as a city authority, these two strategies assist in designing and implementing alterations at programme and institutional levels in such a way that they become more influential as regards urban sector work and policy in general. Concrete measures can be very different in scope. Examples are the inclusion of relevant policymakers and scholars in programmes from the beginning; joint implementation of adaptation projects by different organizations; and the support of national stakeholders in the revision of relevant policies or regulations (such as national road safety directives with risk reduction as an inherent part). From the standpoint of an organization at a national or regional level (such as ministries of urban planning or education), the strategies described in Box 2.10 can 'translate' into the creation of activities to support improved risk reduction integration, knowledge transfer and communication at (and between) different levels. Examples of concrete activities are capacity development programmes for local authorities; the creation of national platforms for researchers, practitioners and policymakers for better knowledge exchange; improved communication (including the selection of the right media, communication strategies and intermediaries if needed); the revision of the curricula of planning education to take risk reduction into account; and making governments and policymakers more involved in universities' research and teaching activities on risk reduction and adaptation (see DFID n.d.; UNSIDR 2011a).

BOX 2.10

Mainstreaming strategies with focus on interorganizational collaboration and capacity development to influence (urban) sector work in general

Strategy V (also called interorganizational mainstreaming for risk govern-ance): promotion of cooperation between existing (urban) actors for capacity development and the harmonization of risk reduction and adaptation within the management of different organizations (working in disaster response, recovery and development).

Strategy VI (also called educational mainstreaming): support for a concep-tual shift in the philosophy that drives sector-specific education in order to allow risk reduction and adaptation to become incorporated into profession-als' (e.g. planners') sphere of activities.

To support capacity development for integrating risk reduction and adapta-tion within one's own organization or other urban institutions is a difficult task. The following are some of the key requisites for successfully develop-ing capacities:

- Understanding of key concepts (as presented in this chapter)
- Assessment of the local context (see Chapter 3 on the city–disasters nexus)
- Assessment of institutional and people's local capacities (see Chapters 4 and 5)
- Ensuring ownership over the capacity development process (of both organizations and citizens)
- Definition of the roles and responsibilities of different urban actors (see Section 4.2)
- Establishment of systems for monitoring, evaluation and learning (see Section 4.2) (adapted from Hagelsteen and Becker 2013).

The different strategies presented in Boxes 2.8–2.10 cannot be seen inde-pendently of each other and thus should not be understood as a step-by-step approach. The different strategies are in fact complementary and reinforce each other. An overview of the characteristics of and differences between the six mainstreaming strategies is shown in Table 2.7. The six

Table 2.7 Overview of strategies for integrating/mainstreaming risk reduction and adaptation into urban organizations' work[a]

Strategy	Focus of change	Aim of change	Illustrative examples
I: Add-on RR and CCA	Activities at local programme level	Adding-on activities focused on RR and CCA	Stand-alone project for early warning (which is not related to the implementing organization's core work)
II: Programmatic mainstreaming	Activities at local programme level	Modification of core work to ensure that it is relevant to the challenges presented by disasters and climate change	Change to more disaster-resistant construction techniques when providing basic infrastructure Provision of back-up systems for infrastructure failure (See Tables 4.1–4.5 for further examples)
III: Organizational mainstreaming	Organizational functioning of implementing body	Institutionalization of RR and CCA to assure its integration at programme level (see Strategies I and II)	Revision of planning tools and corpus of regulations so as to ensure that they consider risk reduction and adaptation Revision of job descriptions of staff to include the responsibility of considering risk and its reduction within daily practice (See Table 4.7 for further examples)
IV: Internal mainstreaming	Organizational functioning of implementing body	Reduction of the implementing body's own risk	Retrofitting of the organization's head office Elaboration of a staff evacuation plan (See Table 4.7 for further examples)

(continued)

Table 2.7 (*continued*)

Strategy	Focus of change	Aim of change	Illustrative examples
V: Interorganizational mainstreaming	(Urban) sector work in general	Improved cooperation and harmonization of RR and CCA to create a functioning (urban) risk governance system and active involvement of at-risk citizens	Establishment of networks for knowledge exchange between different urban actors Online portals providing support for adequate coordination and knowledge sharing (See Table 4.8 for further examples)
VI: Educational mainstreaming	(Urban) sector work in general and education of related professionals in particular	Improved science–policy integration and shift in the philosophy that drives sector-specific education (e.g. planning education)	Establishment of cooperation with local universities for the joint development of projects or operational tools for RR and CCA (See Table 4.8 for further examples)

Note:
a RR: risk reduction, CCA: climate change adaptation.

strategies can be applied by all types of urban actors and can be used in the context of both bottom-up and top-down approaches for urban development planning.

If put into practice, related actions can at all levels be incremental or transformative, i.e. improving either existing risk reduction and adaptation approaches to maintain system functions or provoking systems change for long-term sustainability (see Box 2.3). They aim at achieving a profound and sustainable change in the multilevel institutional framework of urban actors, related policies, decision-taking structures and actions, which ultimately results in the delivery of risk reduction and adaptation measures on the one hand and the building of adaptive capacity on the other hand.

In Chapter 4, some of the most common barriers to successful mainstreaming of disaster risk reduction and adaptation are discussed. They relate to the misconception of the mainstreaming concepts and also to city authorities' organizational setting, including inadequate working structures, control mechanisms, planning basis, political support, human and financial resources and cooperation (Table 4.10). The mainstreaming concept presented in this section considers and addresses these key challenges.

2.5 From urban planning to urban risk governance

The definition of mainstreaming presented in the previous section only becomes meaningful and can be operationalized if it is used in connection with a cross-cutting topic and specific sector work. In this book, the focus is on risk reduction and adaptation as the cross-cutting topics and their integration into urban sector work, with emphasis on urban planning and related urban governance processes.[21] But what do we mean by urban planning? Before continuing to read, ask yourself: how can urban planning be defined? And what kind of actions does it generally involve?

Urban planning is here understood as place-based problem-solving that is aimed at sustainable urban development (Davoudi *et al.* 2009). On this basis, it refers to (a) actions and interventions that are concerned with how urban places and the built fabric are connected with people and nature, and (b) the related processes for ensuring sustainable urban communities, towns and cities (Box 2.11). This involves not only legislative and regulatory frameworks for the adequate development and use of housing, infrastructure, services and land, but also the (inter)institutional, financial and social structures through which such frameworks are implemented, maintained, challenged and transformed. Such a broad understanding of urban planning matches today's reality of city authorities and planners. It further allows use of the terms urban planning, urban development planning, city planning and urban management as synonyms and makes their close interrelation with urban governance processes obvious. Hence, the next question to be asked is: what is urban governance and what does it involve?

While there are many different perspectives and interpretations of the urban governance concept, it generally refers to the multitude of urban actors and related processes that lead to collective binding decisions (which then often become visible in the urban fabric). These different actors and processes together create a 'system of governing' urban development, which is the exercise of coordination and control, in which the state (or government) is not the only and not necessarily the most important actor (see Rhodes 1996; Bulkeley and Betsill 2003).

In accordance with the definition of urban planning given previously, the analysis and improvement of urban governance processes can be understood as being an inherent part of urban planning. It is, however, important to be aware of the fact that there are also some practitioners and scholars who use the urban governance concept to replace the term urban planning. There are two main reasons for this. First, the term urban planning is rather old and historically had a quite narrow definition closely connected to physical master planning. The term urban governance has never had such (outdated) connotations. Second, urban governance expresses more explicitly the increasing recognition of the roles played by different urban stakeholders and at different scales, and the complex interactions between them.

The recent shift towards a governance perspective has taken place not only in the context of urban planning, but also in the field of disaster risk management. Risk governance has become, in fact, a new and prosperous research area. Its focus is on the analysis of systems of governing risk, presuming that there is a complex network of interacting stakeholders with often diverging agendas where no actor is able to control all relevant aspects on its own (see Tehler 2012).

From this perspective, a sustainable mainstreaming of risk reduction and adaptation into urban planning can only be achieved when considering the system of governing urban development *and* the system of governing urban risk. The combined domain of urban risk governance focuses on assessing systems of governing urban risk (i.e. how decisions on urban risk management are taken, communicated and implemented) and, on this basis,

BOX 2.11

Linking urban planning, sustainability and resilience

Urban planning is in this volume understood as place-based problem-solving, which is aimed at sustainable urban development. The importance of achieving sustainable development and global sustainability was established internationally after the Earth Summit in 1992 and strongly influences today's urban planners. The definition of sustainable development that has endured above all others is that put forward by the Brundtland Report: 'Sustainable development is development that meets the needs of the present without compromising the ability of future generations to meet their own needs' (WCED 1987: 43). On this basis, the concepts of sustainable communities and sustainable cities evolved to provide means of reconciling the requirements of sustainable development with the day-to-day business of city authorities and planners (including economic, environmental and social aspects). In practice, there is a great deal of variation in how cities attempt to become sustainable, which is also related to the differing and context-specific sustainability challenge they face.

The issues of disasters and climate change form part of the larger sustainability challenge (IPCC 2007a). Sustainability science has emerged in the twenty-first century as a new academic field seeking solutions to sustainability challenges, and it approaches the world as a complex human–environment system (Kates *et al.* 2001). It addresses the interface between the social and natural sciences, and between science and practice (at global, regional, national and local levels), as well as the task of how to 'operationalize' sustainability. It demands systems thinking, interdisciplinary research and transdisciplinary collaborations (Clark and Dickson 2003; Jerneck *et al.* 2011; Lang *et al.* 2012).

With disasters and climate change being part of the larger sustainability challenge, urban planning can only fulfil its aim of achieving sustainable development if city authorities and other urban actors succeed in mainstreaming risk reduction and adaptation into their everyday practice. Such integrated planning practice leads to urban disaster resilience, which is one important (but not the only) aspect of sustainable urban development (see related Boxes 1.4, 2.2, 2.3 and 3.6).

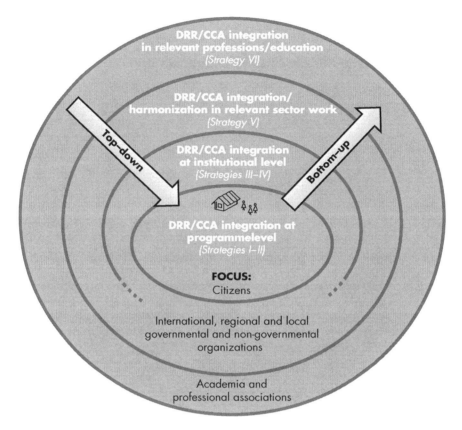

Figure 2.8 Mainstreaming to achieve sustainable (urban) development and risk governance. (See Boxes 2.8–2.10.)

suggests ways in which risk in cities and other urban environments can be better managed by the complex network of interacting stakeholders at different levels. This requires the consideration of:

•   The diversity of urban actors
•   The interactions among them
•   Their respective goals and agendas (see Renn *et al.* 2011).

The mainstreaming strategies described in Section 2.4 were developed on the basis of this understanding. In fact, the combination of the six different strategies is aimed at achieving a profound and sustainable change in the multilevel institutional framework of urban actors, related policies, decision-taking structures and actions. Figure 2.8 links the mainstreaming strategies (defined in Boxes 2.8–2.10) with the governance perspective and related stakeholders. These stakeholders include international, regional and local governmental and non-governmental organizations (e.g. ministries, city authorities, donors, private companies, civil society organizations, professional associations and academia), as well as citizens. If disaster-resilient cities are the aim, all these stakeholders need to be on board and take on their respective roles (see UNISDR 2010a; Wamsler 2009b).

2.6 From risk governance to urban flow management

The recent shift towards a governance perspective to analyse and improve urban risk management (described in Section 2.5) is today accompanied by an increasing interest in the study of urban flows. This is related to the fact that urban flows are a good starting point for the analysis of governing systems as regards their functioning and potential non-functioning when dealing with disturbances (such as disasters and climate change). An example of an urban flow is, for instance, the flow of people or goods on cities' road networks. Such flows can be analysed by asking questions such as: who or what can potentially disturb the flow? And who are the actors that could reduce the risk of such an occurrence? Obviously, the robustness or resilience of a road network is not only influenced by how it is designed. Not only one, but many actors can have both a positive and a negative influence on its functioning and thus would need to be involved to establish and integrate sustainable risk reduction and adaptation processes.

The increasing interest in (the study and management of) urban flows to improve urban resilience underpins the growing recognition that a functioning urban governance system is needed to sustainably reduce risk before, during and after disaster occurrence. Related urban flow studies are, however, mainly concerned with horizontal flows (between cities) and to a lesser extent with vertical flows (such as country–city connections). This relates to the fact that cities are increasingly seen as today's key players in urban flow and risk management. An illustration of this is UNISDR's 'Making Cities Resilient' Campaign (Box 2.12), as well as the 'Ask-Your-Urban-Neighbour' initiative of Training Regions, which is a knowledge exchange tool for European cities (Box 2.13).

Urban flow analysis is also a common theme in urban sustainability science in order to unravel the dynamics and effective management of complex urban systems (Weinstein 2010; see Box 2.11). It is based on a systems approach and the understanding that 'the more complex the system, the

BOX 2.12

The 'Making Cities Resilient' Campaign

The 'Making Cities Resilient: My City is Getting Ready' Campaign was launched in May 2010 to address issues of local governance and urban risk. Based on the five priorities of the Hyogo Framework for Action 2005–2015: Building the Resilience of Nations and Communities to Disasters (see Box 2.7), a ten-point checklist for making cities resilient was developed to which local governments can sign up. By doing so, city authorities commit to implementing disaster risk reduction activities along these ten essentials. As the coordinator of the campaign, UNISDR facilitates regional and global partnerships in support of local governments participating in the campaign, and promotes city-to-city learning and exchanges among cities and local governments.

By Helena Molin Valdes

See also http://www.unisdr.org/campaign/resilientcities

BOX 2.13

Training Regions

Training Regions has the vision to help create sustainable, safe and attractive cities and regions by providing the most qualified arena for global cooperation, networking and the generation of new knowledge. It is a public–private partnership, a triple-helix formation in which governmental authorities, private companies and universities collaborate in reaching the objectives of their respective spheres. Its focus is on training in its widest sense, including resilient urban flow workshops, advice, education, exercises, etc. It supports the development of capacities of cities and regions to protect the well-being of their citizens and to maintain critical societal functions – this by ensuring effective and efficient flows of people, capital, goods and services, regardless of disturbance or disruption, now and in the future.

In 2012, Training Regions launched its 'Ask-Your-Urban-Neighbour' initiative, a knowledge exchange tool for European cities. The idea is to share knowledge with other cities by sharing information regarding how different cities analyse and influence urban flows of people, goods and services.

By Per Becker. See also http://www.trainingregions.com
or http://www.resilientregions.org

greater the risk of systemic breakdown, but also the greater the potential opportunity' (WEF 2012: 8). Approaching cities from a systems perspective stands in contrast to former approaches, which have focused on the analysis of single organizations to push forward sector-specific specialization and improvements in technical performances. City officials and planners still tend to focus on such traditional approaches (Troedsson 2011). The different strategies for mainstreaming risk reduction and adaptation described in Section 2.5 can assist city authorities and other urban actors in gaining a wider perspective and approach to deal with increasing disasters and climate change. They match and support the ideas that underlie urban flow analysis and management and related systems thinking to achieve urban resilience.

2.7 Summary – for action taking

There are no universally accepted definitions of many of the concepts that are central to the understanding of urban risk reduction and adaptation. Independent of the actual terms used, it is, however, crucial for city authorities and other urban actors to possess a coherent conceptual framework that can guide their daily practice and ensure the comprehensive management of increasing disasters and changing risk patterns.

The ideal of urban resilience can be understood as cities and other urban environments that can easily live and cope with an ever-changing, and sometimes risky, environment, including climate- and non-climate-related, small and large-scale disasters. It is the desire to restore the historical function of cities as places where inhabitants can find safety and protection (see Boxes 1.4, 2.2 and 2.11).

Translated into more operational concepts, a disaster-resilient city can be understood as a city that has successfully managed to implement or support measures to strengthen the capacity of individuals, households, communities and institutions to:

1. Reduce or avoid current and future hazard exposure
2. Reduce current and future susceptibility to hazards (to increase the ability of locations to withstand hazards)
3. Establish functioning and flexible mechanisms and structures for disaster response
4. Establish functioning and flexible mechanisms and structures for disaster recovery (see Box 2.4 and Figure 2.6).

These risk reduction and adaptation measures correspond to the factors that cause disaster risk and ultimately disasters. The first measure listed addresses the hazards to which an area (or another system) is exposed; the following three measures address the factors that influence the vulnerability of an area (or another system) when facing these hazards (Figure 2.9). All four measures need to be based on continuous risk assessments, monitoring and learning mechanisms, which place at least the same importance on vulnerability issues as on hazard patterns.

Furthermore, resilient cities can only be achieved if city authorities and other urban actors have succeeded in sustainably mainstreaming risk reduction and adaptation into urban planning so that these issues become institutionalized at all levels. The misunderstanding of the concept of mainstreaming is one of the key barriers to achieving sustainable risk reduction and adaptation planning. In order to mainstream risk reduction and adaptation, urban actors would need to revise their day-to-day work and their organizations' management and functioning in order to:

- Ensure that their programme activities do not increase risk and, if possible, reduce risk
- Institutionalize risk reduction and adaptation so that its integration at programme level becomes a standard procedure (including procedures for anticipation, monitoring, actual adaptation and learning)
- Assure their organizations' own functioning during times of disasters and climate change
- Cooperate with others to create a functioning multilevel system of governing risk
- If possible, push forward better science–policy integration and improved education on urban resilience (see Boxes 2.8–2.10 and Figure 2.8).

On this basis, disaster resilience can be understood as a process that can be assessed for (a) the extent to which city authorities and other urban actors are successful in implementing measures for (improving capacities for) reducing current and future risk, and (b) the degree to which they succeed in mainstreaming risk reduction and adaptation. Related efforts are an expression of existing adaptive capacities and become 'visible' in urban actors that are (more or less) able to reduce and adapt to evolving and changing risk in a flexible, dynamic and effective manner. In line with this, urban risk reduction and adaptation include not only the improvement of current approaches to maintain systems functions, but also the transformation of systems for achieving long-term sustainability. The increasingly prominent concept of sustainable urban transformation can thus be understood as the denotation

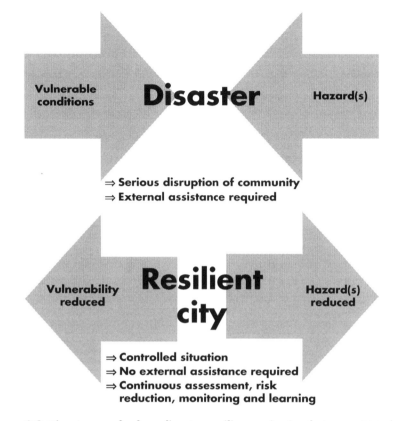

Figure 2.9 Disasters and urban disaster resilience – in simple terms. Note that in this simplified illustration, deficiencies in mechanisms and structures for disaster response and recovery are merged under the notion of vulnerable conditions/vulnerability.

of the recent advances made in the understanding of risk reduction and adaptation, which today incorporate the notion of sustainability and transformation (as described in detail in this chapter and further advanced in Chapter 6).

Cities that are not disaster resilient and that have low adaptive capacity are prone to long-term suffering after hazard impacts (including climatic extremes and variability) and are then more prone to adverse impacts from even smaller hazard events. In contrast, resilient cities may be in an even stronger position post-event. This may be due to preparedness measures that have built up institutional adaptive capacities, which allow the effective use of the recovery phase (and related financing) for risk reduction and adaptation and, thus, make it possible to 'build back better'.

2.8 Test yourself – or others

? The following questions can be used to test yourself or others. The answers to most questions can be found explicitly in Chapter 2.

1. Why does a particular hazard event affect some city dwellers (or cities) more than others?
2. How do the concepts of risk reduction, adaptation, response, recovery and development relate to each other?
3. What are the differences and similarities between disaster risk reduction and climate change adaptation?
4. What is risk assessment, what does it include and why is it so important for achieving sustainable risk reduction and adaptation?
5. What kind of measures can be implemented to reduce the risk of flood? How do the measures you propose address different risk factors? Do the proposed measures also increase city dwellers' capacity to reduce risk?
6. What is adaptive capacity? Give examples of how adaptive capacity can be increased at different levels (i.e. of individuals, communities and urban institutions).
7. What is the difference between integrating and mainstreaming risk reduction into urban planning practice?
8. What are the different potential strategies for mainstreaming risk reduction and adaptation into urban planning practice? Do they have to be implemented one after the other?
9. How can climate change affect urban risk as well as urban communities' capacities for risk reduction?
10. The two expressions 'most at risk' or 'most vulnerable' (e.g. to describe a specific urban community) might express just the same or different things. Please explain why this is the case and what the differences could be.
11. Give examples of two types of urban flows and how their analysis could lead to improved risk reduction and adaptation. On this basis, describe how urban flow thinking is different from traditional approaches to fostering urban resilience.
12. What does the concept of urban governance entail? How is the role of the state or local government perceived in this concept?

Case studies are also a good basis for assessing the knowledge presented in this chapter.

✎ **Test yourself scenario 2.1: key concepts central to the understanding of risk reduction and adaptation.** The Minister of Housing and Planning has called a meeting to discuss the need and possibilities for improving the existing capacities for urban risk reduction and adaptation. You are part of an expert team, which has been asked to support the Ministry in its efforts. The Minister has invited a large and diverse group of agencies: various ministries, local government representatives, the Water Utilities Corporation, the National Red Cross Society and various national NGOs. The objective of having this meeting is to kick off a process for future interinstitutional cooperation with the aim of increasing existing capacities for urban risk reduction and adaptation. In preparation for the meeting, the Minister asks you to write a simple and short briefing note that should give all the different participants a common understanding on some key concepts and help avoid any misunderstandings during the meeting. In addition to preparing the briefing note, the Minister asks you to prepare a presentation of a maximum of 10 minutes to summarize the briefing note at the beginning of the meeting.

 Test yourself scenario 2.2: linking risk reduction, adaptation and development. You are working for an international organization as the focal point for urban risk reduction and adaptation. A new staff member approaches you, asking you to clarify the differences between some of the terms he heard during the last staff meeting and which he had scribbled on a piece of paper:

Humanitarian aid, Relief & rescue, Recovery, Reconstruction, Rehabilitation, Development assistance, Technical cooperation, Disaster risk reduction (DRR), Climate change adaptation (CCA), Early recovery, Development-oriented response.

Task: Please categorize the different terms and then discuss and answer the following questions:

1. Which of the terms mean (more or less) the same?
2. What is the difference between the first seven terms and the four last terms?
3. Which of the terms form part of disaster risk management?
4. Which of the terms are linked to disaster risk reduction? How?
5. What is the difference between the first seven terms and the following two terms (i.e. DRR and CCA)?

Test yourself scenarios 2.3a and 2.3b: risk and how to address different risk components.

2.3a. You are an urban planner and risk reduction consultant on your way to visit the Imaginato community in Mexico, which was affected by a recent

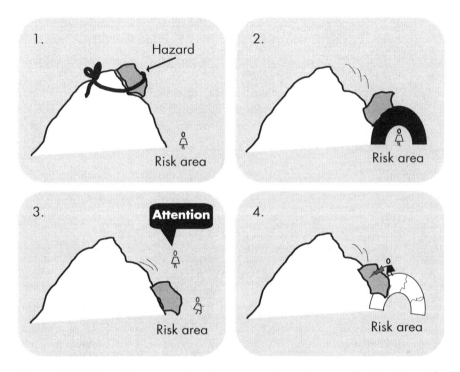

Figure 2.10 Potential approaches/measures to reduce and adapt to risk.
(See Box 2.4.)

landslide. You pass houses made of local wood, partly buried under mud. The hilly terrain makes access to the community difficult, but finally you arrive. The community leader Alejandro welcomes you. Jorge has also been waiting for you. He is a staff member of the NGO you are working for, which has operated in the area for the past two decades. Alejandro and Jorge present you to the local committee and two other staff members of the NGO. Alejandro opens the meeting with the words: 'Every year, more and more landslides occur in our community. We finally want to address this issue more systematically. We are very pleased that you have come to help us understand the different options we have to reduce our risk. But let's first walk together to the community centre where we can sit together more comfortably.' Walking to the community centre you have some minutes to think about the best way to summarize and present, in a comprehensive way, the different types of approaches that can be taken to reduce risk (from landslides). You bear in mind, and quickly recapitulate, the understanding of risk. What are the key factors that contribute to risk? How can they be addressed? How can this theoretical knowledge be best illustrated to make everybody understand? And what could be a good concrete example to illustrate each type of measure?

2.3b. The community committee is happy with your presentation on the different approaches they can take to reduce risk and at the community centre Jorge is fixing your illustrations on the wall (Figure 2.10). During the lunch break, he asks you to assist him in understanding the aspects that it is important to assess in order to design concrete programme activities. In order to do so, you select a specific risk reduction activity (such as the provision of an emergency shelter) in order to discuss with him different alternatives for designing this activity most effectively. In this context, please go through the following questions:

1. How could existing local coping strategies (i.e. used capacities) be taken into account?
2. How could other unused capacities of the beneficiaries be taken into account?
3. Could there be a need to add on new activities/aspects (in case of unsustainable coping strategies or low-capacity households)?
4. Should the activity (only) aim at directly reducing a specific risk factor or (also) provide related capacity?
5. Could the activity be designed in such a way as to reduce different risk factors at the same time?
6. How could the activity be designed in order to increase people's level of participation?
7. Finally, discuss how the selected activity could be further improved in order to increase the probability of it being implemented and sustained, using the following aspects and criteria:

    • Social acceptance
    • Technical feasibility and maintenance
    • Administrative issues (e.g. organizational structures, staffing)
    • Political and public support
    • Legal viability (e.g. existing local and state authorities, and regulations)
    • Economic issues (e.g. cost-effectiveness, available funding)
    • Environmental issues (long-term versus short-term benefits).

Please note that the case study is not meant to be a comparison of different activities. A comparison would also have to take into account the expected (long-term) impacts of the different alternatives in terms of their potential to reduce the number of deaths, injuries and property and economic losses.

2.9 Guide to further reading

On risk and risk reduction

Coppola, D., 2011. *Introduction to international disaster management.* 2nd ed. London: Elsevier.

ISO (International Standards Office), 2009. *ISO 31000 Risk management: principles and guidelines.* 1st ed. Geneva: ISO.

Twigg, J., 2004. *Disaster risk reduction: mitigation and preparedness in development and emergency programming.* No. 9, Good Practice Review. London: ODI.

UNISDR (United Nations Office for Disaster Risk Reduction), 2009. *Terminology: disaster risk reduction.* Geneva: UNISDR.

Wisner, B., *et al.*, 2004. *At risk: natural hazards, people's vulnerability and disasters.* 2nd ed. Abingdon and New York: Routledge.

On adaptive capacities

Wamsler, C., Brink, E., and Rantala, O., 2012. Climate change, adaptation, and formal education: the role of schooling for increasing societies' adaptive capacities in El Salvador and Brazil. *Ecology and Society*, 17 (2), Art. 2 [special issue on 'Education and differential vulnerability to natural disasters'].

On mainstreaming risk reduction and adaptation

Benson, C., Twigg, J., and Rossetto, T., 2007. *Tools for mainstreaming disaster risk reduction: guidance notes for development organizations.* Geneva: ProVention Consortium.

Wamsler, C., 2009a. *Operational framework for integrating risk reduction and climate change adaptation into urban development.* Brookes World Poverty Institute (BWPI), Working Paper Series No. 101 & Global Urban Research Centre (GURC), Working Paper Series No. 3. Manchester, UK: BWPI/GURC.

On risk reduction and adaptation

Hamza, M., Smith, D., and Vivekananda, J., 2012. *Difficult environments: bridging concepts and practice for low carbon climate resilient development.* Brighton, UK: IDS Learning Hub.

UN-IATF/DR (2006). *On better terms: a glance at key climate change and disaster risk reduction concepts.* Geneva: United Nations Working Group on Climate Change and Disaster Risk Reduction of the Inter-Agency Task Force on Disaster Reduction (UN-IATF/DR).

UNISDR (United Nations Office for Disaster Risk Reduction), 2008a. *Briefing note 01: climate change and disaster risk reduction.* Geneva: UNISDR.

On adaptation, resilience and transformation

Cannon, T., and Müller-Mahn, D., 2010. Vulnerability, resilience and development discourses in context of climate change. *Natural Hazards,* 55 (3), 621–635.

IPCC (Intergovernmental Panel on Climate Change), 2012a. *Managing the risks of extreme events and disasters to advance climate change adaptation (SREX).* A Special Report of Working Groups I and II of the Intergovernmental Panel on Climate Change [C. Field *et al.*, eds.]. Cambridge: Cambridge University Press.

Pelling, M., 2012. *Adaptation to climate change: from resilience to transformation.* Abingdon and New York: Routledge.

Plan International, 2012. *Climate extreme: how young people can respond to disasters in a changing world.* Plan International and Children in a Changing Climate. (Child-friendly version of the SREX report, see IPCC 2012a.)

See also recommended readings included at the end of the other book chapters.

2.10 Web resources

Emergency Events Database (EM-DAT). EM-DAT, by the Centre for Research on the Epidemiology of Disasters (CRED), keeps global data on the occurrence and effects of mass disasters from 1900 to the present: http://www.emdat.be

Gender and Disaster Network (GDN) (see Box 2.14). The GDN is an educational project focusing on gender relations in disaster contexts: http://www.gdnonline.org

Global Network for Disaster Reduction (GNDR) (see Box 2.14). A global network of civil society organizations working together to improve disaster risk reduction policy and practice at the decision-making level: http://www.globalnetwork-dr.org/home.html

Home of Radical Interpretations of Disaster and Radical Solutions (RADIX) (Box 2.14). RADIX is a home for discussion, working papers, opinion pieces, resources and links, which can lead to the development of radical (as in fundamental or concerning root causes) interpretations of and radical solutions for disasters worldwide: http://www.radixonline.org

Intergovernmental Panel on Climate Change (IPCC). Established in 1988 as a scientific intergovernmental body set up by the World Meteorological Organization (WMO) and the United Nations Environment Programme (UNEP), the IPCC provides decision-makers and others interested in climate change with an objective source of information about the issue: http://www.ipcc.ch

BOX 2.14

RADIX and other networks for the critical urban planner

The UN International Decade for Natural Disaster Reduction (IDNDR), 1990–1999, had come and gone. What changes had all the science and conferencing accomplished on the ground? Within the first months after the IDNDR, Maureen Fordham and Ben Wisner saw the avoidable destructions after earthquakes in India and El Salvador. They decided that a more radical approach to disaster reduction is needed and founded RADIX (http://www.radixonline.org) as a hub for material that deals with root causes.

The 'mother site' of RADIX was Gender and Disaster Network (GDN) (http://www.gdnonline.org), which provides considerable amounts of very practical advice, case studies and methods. Gender is more than a 'risk factor'. The positive and proactive role of women and girls in planning is seldom recognized and hardly realized in its full potential. GDN had emerged from a meeting during the July 1997 Natural Hazards Research and Applications Center workshop in Boulder, Colorado. GDN documents and analyses women's and men's experiences before, during and after disaster, situating gender relations in broad political, economic, historical and cultural context. It works across disciplinary and organizational boundaries and fosters information sharing and resource building among network members

Radix and GDN are both members of a larger network called GNDR (Global Network of Civil Society Organizations for Disaster Reduction) (http://www.globalnetwork-dr.org/home.html). It is a major international network of non-governmental and not-for-profit organizations committed to working together to improve the lives of people affected by disasters worldwide. Members compile evidence for investment at the local scale, where communities, civil society and local government have untapped potential for risk reduction and enhancement of livelihoods. The website is a gateway to its large survey, methods and many case studies.

By Maureen Fordham and Ben Wisner

International Recovery Platform (IRP). Conceived at the World Conference on Disaster Reduction in Kobe, Hyogo, Japan in 2005, the IRP is a thematic platform and key pillar for the implementation of the Hyogo Framework for Action (HFA) (see Box 2.7). Its main functions are to identify gaps and constraints experienced in post-disaster recovery and to catalyse the development of tools, resources and capacity for resilient recovery: http://www.recoveryplatform.org

Training Regions (see Box 2.13). Training Regions is an urban flow management arena where different stakeholders can meet to promote the development of a more robust society, especially within cities and regions. Note that during the publication process of this book Training Regions changed its name to Resilient Regions Association: http://www.trainingregions.com and/or http://www.resilientregions.org

PreventionWeb. PreventionWeb is a project of UNISDR launched in 2007 to serve the information needs of the disaster risk reduction community: http://www.unisdr.org/we/inform/preventionweb

United Nations Office for Disaster Risk Reduction (UNISDR). UNISDR and its campaign 'Making Cities Resilient' (see Boxes 2.7 and 2.12): http://www.unisdr.org, http://www.unisdr.org/campaign/resilientcities

See also web resources included at the end of the other book chapters.

3 The city–disasters nexus: a two-way relationship

Learning objectives

- To explore the reciprocal linkages between disasters and cities

  ○ To analyse how disasters and climate change impact cities
  ○ To identify cities' characteristic features and how they influence risk

- To discuss the reciprocal linkages between disasters and urban planning processes – by considering the life cycle of disasters from causes, to short- and long-term impacts, to post-disaster response and recovery
- To identify the complex and manifold connecting points between urban planning, risk reduction and adaptation

In-depth knowledge of the interlinkages between disasters and cities is an indispensable precondition for mainstreaming risk reduction and adaptation into urban sector work; it allows city authorities and other urban actors to modify their work in order to address increasing risk and disasters (see Section 2.4). So the question to be asked is: how are disasters and cities interlinked?

The perception that disasters are the uncontrollable cause, and the destruction of the built environment is the effect, is widespread (Bosher 2008). Consequently, planning responses are often very limited, solely focusing on constructive aspects and mainly on the post-disaster context. The reality is, however, more complex than this simple one-way, cause-and-effect relationship. The city–disasters nexus is, in effect, a reciprocal and multifaceted relationship including:

- The effects of disasters and climate change on cities, and the reverse interrelation; that is:
- The influence of cities' characteristic features on (increasing) risk and disaster occurrence.

In order to analyse the reciprocal relationship between cities, disasters and climate change in a systematic and comprehensive way, urban actors first and foremost need an understanding of what makes urban areas different from rural areas. The importance of understanding urban–rural differences can be illustrated using the example of the Haiti earthquake in 2010 and the following statement from Esteban Leon, UNHABITAT:

> I remember how I deployed to Haiti in 2010, a few days after the earthquake, and the chaotic situation I lived in there. It is very difficult to describe, except to note that of all of the agencies that responded to that earthquake, only a very small handful had any urban expertise or understanding of the urban context, and consequently no urban framework for engaging [with the situation]. That, of course, did not help respond in a coherent and coordinated manner.[1]

But what are the characteristic and distinctive features of cities?

Put simply, the process of urbanization brings about profound physical, environmental, socio-cultural, economic and political/institutional changes, which mean that urban areas are unlike rural areas, being characterized by distinctive physical, environmental, socio-cultural, economic and political/institutional features. These city features are described in the following section.[2]

3.1 City features

Urbanization and the resultant changes find their visible expression in the so-called urban fabric. The urban fabric is characterized by distinctive physical features, which relate to aspects such as population density, land coverage and vegetation, architectural details, infrastructure, organization of structures in space and the relationship between buildings and topographic aspects (Table 3.1).

Many of the environmental, socio-cultural and economic changes that make urban areas different from rural areas can be attributed to the physical features of the urban fabric listed in Table 3.1. This becomes especially obvious when looking at environmental changes. These changes are manifested in the urban ecosystem, which is characterized by distinctive environmental features as regards precipitation, wind, temperature, air quality, humidity, solar radiation, soil, water bodies, flora, fauna, noise, waste and wastewater (Table 3.2).

Most of the environmental features presented in Table 3.2 are a direct result of the urban fabric's physical characteristics. This relates to the progressive

Table 3.1 Distinctive physical features of cities: the urban fabric

Physical features	Distinctive urban characteristics[a]
Population density	Population density and overpopulation (increased)
	Access to marginal areas (reduced)
Land coverage and vegetation	Built-up surface area (increased)
	Size, location and distribution of green and recreational areas (reduced)
	Tree coverage (reduced)
	Access to affordable space (reduced)
	'Consumption' of land (including rural land) (increased)
Architectural details	Height of buildings (increased)
	Differences in the height of buildings (more varied)
	Construction materials and colours (different; more influential e.g. of streets)
	Construction techniques (less traditional, more advanced)
	Shape of buildings (more varied; eventually more restricted)
Infrastructure	Infrastructure network and connectivity (increased; more congested)
	Dependency on infrastructure network (increased)
	Flows (e.g. material and people) (increased)
Organization of structures in space	Distance between buildings (reduced)
	Concentration and interdependence of buildings, services and infrastructure (increased)
	Concentration and interdependence of political and economic centres (increased)
	Orientation of buildings (more restricted)
	Street layout and street orientation (denser; more restricted)
Relationship between buildings and topographic aspects	Proximity to large bodies of water (reduced)
	Relation to nearby hills and valleys (more difficult to account for)
	Sloping terrain (more difficult to account for)

Note:
[a] The described characteristics and related connotation (i.e. increased or decreased) are only an indication and are thus not applicable to every city.

sealing of, and construction on, green areas together with cities' high population densities, which amongst other things bring about increased energy use, emissions and heat. This last, known as the 'heat island effect', is caused by heat storage in and radiation from the built fabric, and the high amount of outlet air, for instance from heating, industrial processes and traffic.

The first six environmental factors listed in Table 3.2 are so-called abiotic ecological factors, which make up the typical urban climate. In contrast to the surrounding areas, the urban climate is generally wetter, less windy, hotter, more polluted, less humid and cloudier (Adam 1988; RGS 2010).

Table 3.2 Distinctive environmental features of cities: the urban ecosystem and climate

Environmental features	Distinctive urban characteristics[a]
Precipitation	Rainfall (increased)
	Snowfall (reduced)
Wind	Wind speed and exchange (reduced)
	Local wind circulation, gusts and turbulence (increased)
Temperature	Temperature (increased)
Air quality	Emissions (increased)
	Dust particles (increased)
Humidity	Air humidity (reduced)
	Fog and cloudiness (increased)
	Evaporation (reduced)
Solar radiation	Length/amount of natural lighting (reduced)
	Intensity of solar radiation (reduced)
Soil	Ground sealing and compression (increased)
	Soil quality (reduced)
Water bodies	Ground water level (reduced)
	Ground water quality (reduced)
	Surface water quality (reduced)
	Water flows (more regulated)
	Dependency on other (rural) areas for ecosystem services (increased)
Flora	Vegetation cover (reduced)
	Biodiversity of vegetation (reduced; specific city vegetation)
	Growing season (increased)
	Vegetation forms (specific forms developed on facades, roofs, etc.)
	Dependency on other (rural) areas for ecosystem services (increased)
Fauna	Biodiversity of species (generally reduced; overpopulation of some species)
Noise	Noise (increased)
Waste and wastewater	Amount of waste (increased)
	Type of waste (more hazardous)
	Wastewater (increased; and increased mix of rainwater and backwater)

Note:
[a] The described characteristics and related connotation (i.e. increased or decreased) are only an indication and are thus not applicable to every city.

However, there are strong inner-city differences which are, for instance, caused by strong localized wind currents and turbulence.

After having looked into the physical and environmental features of cities, let's discuss other aspects that make cities different from rural environments. Before continuing to read, ask yourself: what are the characteristic socio-cultural, economic and political/institutional features of cities? And how are they related to the physical aspects of the urban fabric described above?

At first sight, the socio-cultural features that characterize urban areas are less connected to the physical changes caused by urbanization. They manifest in a characteristic urban society and culture (or so-called urban life), which has distinctive features related to family structures, social cohesion, social inequality, public participation, values and habits, population diversity, health and security aspects (Table 3.3).

A more thorough analysis shows, however, that several of the socio-cultural aspects of urban life (listed in Table 3.3) are also directly linked to the physical features of the urban fabric. High population densities,

Table 3.3 Distinctive socio-cultural features of cities: urban society and culture

Socio-cultural features	Distinctive urban characteristics[a]
Family structures	Nuclear family structures (increased)
	Extended family structures (reduced)
	Female-headed households (increased/reduced)
Social cohesion	Sense of community (reduced)
	Sense of family and family obligations (reduced)
	Local leadership structures (reduced)
	Social interaction (increased)
	Anonymity (increased)
Social inequality	Segregation of different population groups (increased)
	Gender equality (increased)
Public participation	Public participation (reduced)
Values and habits	New value systems (increased individualism, etc.)
	Traditional/indigenous knowledge (reduced)
	Secularization (increased)
	Consumer perspective (increased, e.g. increased consumption of meat, fast food, energy)
	Formal education (increased)
	Illiteracy (reduced)
Diversity of people	Diversity of people/heterogeneous communities (also due to rural migrations) (increased)
Health and security	Health status (generally increased; certain health problems may increase such as obesity or HIV/AIDS)
	Violence and organized crime (increased)

Note:
[a] The described characteristics and related connotation (i.e. increased or decreased) are only an indication and are thus not applicable to every city.

overpopulation, the lack of affordable space and the lack of green and recreational areas can, for instance, influence family structures, social cohesion and sense of community. In overcrowded conditions, issues such as competition for space and poor infrastructure (with, for instance, lack of or leaking wastewater pipes) can easily generate conflicts between neighbours. Likewise, the lack of infrastructure or its failure to provide adequate water, sanitation, drainage, roads and footpaths can increase the health burden and the workload of residents, especially for women (IFRC 2010; Tacoli 2012; UNHABITAT 2010b; see Box 2.14). In fact, women and girls in informal settlements are often expected to spend long hours fetching water or staying at home to care for family members made sick by poor-quality water and inadequate sanitation – unpaid work which can reduce their educational and employment opportunities (UNHABITAT 2010b). Even where there is piped water but it is, for instance, only accessible during short periods, without residents knowing when, women have little opportunity to spend time away from their homes or combine their household responsibilities with having a job. In addition, the often poor accessibility of certain urban areas, together with the lack of nearby public leisure space, can isolate certain groups (such as the elderly and women with small children) and make them even more 'bound' to their compact homes. Some features of the urban fabric can also be linked to the occurrence of gender-based violence and the transmission of sexually transmitted diseases. HIV prevalence among urban women in sub-Saharan Africa is, for instance, already much higher than among rural women (UNHABITAT 2010b).[3] Houses that lack water and sanitation force girls and women to seek toilets or washing areas away from their homes, also after nightfall, which can put them at higher risk of sexual assault and violence. Furthermore, for women and children living in the periphery of cities, long working days in the city become even longer and more dangerous in the absence of adequate infrastructure and transport systems.[4]

Finally, the economic and political/institutional changes caused by urbanization processes manifest themselves in cities' characteristic urban economies and governance systems. In contrast to more rural environments, these systems have distinctive features with respect to people's incomes, prices, livelihood practices, resource availability, the presence and concentration of certain institutions, related control power, public expectations and public reliance on institutions and social security systems (Table 3.4).

The interconnections with the urban fabric are manifold. Space restrictions, for instance, make it impossible for city dwellers to be self-sufficient or rely upon agriculture as their main income source. Likewise, growing food in combination with another job, as a diversification strategy, is generally

Table 3.4 Economic and political/institutional features of cities: the urban economy and governance system

Economic and political features	Distinctive urban characteristics[a]
Agricultural versus non-agricultural incomes	Agricultural versus non-agricultural incomes (reduced)
Subsistence versus money economy	Subsistence versus money economy (reduced) Dependency on (rural) food market (increased) Monetary income (increased) Prices for goods and services (increased) Income and employment opportunities (increased/less equal) Labour as critical asset (increased)
Urban livelihood practices	Specialization versus diversification (increased; related overexploitation of natural assets increased) Working space (reduced) Use of housing as a productive asset (increased)
Resource availability	Resource availability (increased) Resource distribution (less equal)
Institutions	Presence and concentration of different institutions and access to their services (increased)
Control power	Control of compliance with legal frameworks (increased)
Public expectations	Public expectations (increased). This includes, for instance, the provision of infrastructure, services, employment and measures for risk reduction and adaptation
Public reliance on institutions and social security systems	Public reliance on social security and welfare systems (increased) Public reliance on institutions, urban governance systems and related services (increased)

Note:
[a] The described characteristics and related connotation (i.e. increased or decreased) are only an indication and are thus not applicable to every city.

not a viable option in cities. In many low-income settlements, people draw their livelihoods from household-based small businesses or service jobs for others living close by. If hazards impact the urban fabric, this situation translates directly into reduced income and, possibly, even the loss of people's income source. In cities, the urban commodity market makes urban dwellers (in contrast to rural dwellers) generally more dependent on money to access resources like food, water and building material, which also increases the importance of labour as a critical asset, especially for the urban poor (Pelling 2003). Consequently, the loss of labour power (e.g. due to reduced demand for workers after disasters, illness or injury) can disproportionally affect the urban poor. Whilst the spatial concentration of political and economic centres translates, generally, into more resources and jobs, albeit not equally accessible, it also leads to increased expectations and reliance on institutions, social security and welfare systems. Many urban dwellers in fact rely on public authorities to solve their problems (including extreme climatic events and variability), rather than taking any risk reduction or adaptation actions themselves (see Section 5.5). Furthermore, cities host specifically dependent groups, such as the large number of urban migrants who are living far away from inherent social safety nets such as the (extended) family. Finally, cities not only concentrate political and economic institutions, they also present a spatial concentration of educational bodies, research institutes, civil society organizations, as well as religious or spiritual centres, which offer their services to citizens.

In sum, urbanization leads to a very characteristic urban fabric, ecosystem, climate, society, culture, economy and governance system in which the cities' different physical, environmental, socio-cultural, economic and political/institutional features influence each other. A comprehensive analysis of the city–disasters nexus needs to take all of these factors and related urban–rural linkages into account, and is an indispensable precondition for identifying adequate measures for urban risk reduction and adaptation.

3.2 Cities' influence on disasters

This section analyses the influence of cities on disaster occurrence. More specifically, it investigates how the urban fabric and related planning practice influence disasters and risk, including climatic extremes and variability.[5] But how can this influence be analysed in a comprehensive way? Before continuing to read, try to answer this question by quickly recapitulating the concepts of risk and disaster described in Section 2.1.

In order to investigate cities' influence on disaster occurrence, the urban fabric's characteristic physical and related environmental, socio-cultural, economic and political/institutional features need to be linked to the four different factors that influence disaster risk. As described in Section 2.1, the risk level of a specific area is influenced by current and future hazards, location-specific vulnerabilities and the functioning of associated mechanisms and structures for disaster response and recovery (Figure 3.1). On the basis of this analytical framework, let's start with the first risk factor. Before continuing to read, ask yourself: how can the urban fabric and related planning practice influence existing and future hazards?

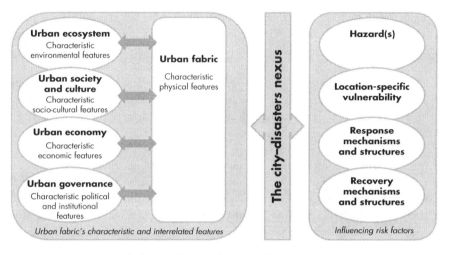

Figure 3.1 Framework for analysing the city–disasters nexus (from an urban planning perspective).

There are many different ways in which the urban fabric can exacerbate existing and future hazards and ultimately increase the number of disasters. The urban fabric has an influence on both the characteristics and the occurrence of hazards. Related key aspects are described in the following text and are summarized in Box 3.1.

First, the urban fabric can intensify existing hazards through its direct influence on the urban climate that results, amongst other things, in increased rainfall, higher temperatures, local wind circulations, turbulence and gusts (see Box 3.1, no. 1 and Table 3.2). Related inner-city differences are strong, both vertically and horizontally. For example, urban flash floods may only affect a few streets in a community (see Table 2.2). In the case of heat, people living under the roof or on the ground floor of multistorey buildings are more exposed than other inhabitants. Even the increase in urban droughts can in part be linked to the characteristic features of the urban fabric. High population densities, the high proportion of surface sealing, little shadowing and heat emissions from, for instance, transport and industries contribute to the urban heat island effect. This situation is aggravated by the high water withdrawal for use in cities due to the high concentration of businesses, industries and people, and many city dwellers' lack of understanding about the effective and responsible use of water. On a more positive note, reduced average wind speed and reduced snowfall, as well as increased temperatures, might become positive urban features during windstorms or cold-waves, respectively.

Second, the urban fabric can create new hazards (see Box 3.1, no. 2). Influencing factors include architectural details and the organization of structures in space. For example, fire can be caused by construction or architectural aspects, such as unsafe electrical connections, or antennas and electrical equipment on top of buildings that attract lightning (Figure 3.2). Landslides can be a direct consequence of the organization of structures in open space, for example, building on watersheds that modify hydraulic regimes and destabilize slopes. Other causes include the destabilization of slopes through the colonization of garbage landfills, deforestation or inadequate construction techniques. The creation of new hazards can further be related to the linkages between the urban fabric and socio-economic aspects, such as urban livelihood practices. As an example, (formal and informal) production processes in densely built settlements can cause 'industrial' explosions leading to fire, whilst related overexploitation of natural resources can also trigger landslides.

The third way in which the urban fabric can exacerbate hazards relates to the expansion of the built fabric into hazard-prone areas, thus increasing exposure (see Box 3.1, no. 3 and Figure 3.3). High population density, in

Figure 3.2 Electrical connections in Rocinha, Rio de Janeiro, Brazil. Source: Christine Wamsler.

Figure 3.3 Houses in Ulm, Germany, built 'over' a tributary stream of the river Danube. Source: Christine Wamsler.

combination with a lack of adequate (or affordable) inner-city land and related socio-economic issues, can lead to the expansion of the urban fabric into areas that are exposed to natural hazards (for instance, near to rivers or on steep slopes) and closer to other types of hazards (such as industries, toxic disposal sites or dangerous streets). The expansion of the built fabric into hazard-prone areas is related to manifold urban–rural linkages. As cities grow, they 'consume' rural land around them (irreversibly transforming rural lives), whilst rural dwellers who commute or migrate to cities are often the ones who end up settling on hazardous urban or peri-urban land. The expansion of the urban fabric into hazard-prone areas is taking place at the city scale, as well as at the local household level where lack of space means that building on slopes, close to landfills or other potential hazards cannot be avoided. This situation, together with the close proximity of land use for residential, industrial and transport purposes, also strongly increases the possibility of compound hazards, i.e. the simultaneous occurrence of different hazards, which is no. 4 in Box 3.1.

Furthermore, the urban fabric intensifies or creates new hazards through its high level of emissions (Box 3.1, no. 5). In addition to directly adding to increased heat and air contamination, high emissions in cities also indirectly create new hazards through their contribution to climate change. Cities,

---

BOX 3.1

How the urban fabric can exacerbate hazards – and ultimately increase disaster occurrence

1. Intensification of existing hazards through the urban fabric's influence on the urban climate (resulting, for instance, in increased rainfall, temperature, local wind circulation, turbulence and gusts).
2. Direct creation of new hazards, mainly fire and landslides, through architectural details, organization of structures in space and urban livelihood practices.
3. Expansion of the urban fabric into hazard-prone areas, thus increasing hazard exposure.
4. Increased potential for compound hazards, mainly due to the close proximity of different types of land use intended for physical, social and economic functions of the city (e.g. residential, industrial and transport purposes).
5. Indirect creation of new hazards through high emissions, thereby fostering climate change and reshaping climate-related hazard occurrence both locally and globally (e.g. increased rain- and snowfall, wind speed, temperature and heatwaves, as well as earthquakes due to the melting of permafrost).
6. Constantly changing hazard patterns due to dynamic urbanization processes.

whilst covering only about 2 per cent of the Earth's surface, are said to produce the vast majority (around 75 per cent) of the world's carbon dioxide emissions (UNHABITAT 2011) (see Box 1.2 and Section 3.4 for a more detailed discussion of this issue). Associated climate change reshapes hazard occurrence, including patterns of precipitation, wind and temperature, which is likely to outweigh the potential positive effects of the urban climate (such as reductions in snowfall and overall wind speed or higher winter temperatures [see Tables 2.2 and 3.2]). In addition, even earthquakes, which are generally considered to be non-climate-related hazards, can be increased when caused by the melting of permafrost (see Table 2.3). On a more positive note, the dense layout of the urban fabric means that a relatively high number of people can easily be provided with infrastructure and services, which can lead to reduced emissions per person compared with rural and remote areas.

Finally, the dynamic urbanization processes continuously reshape the urban fabric, leading to constantly changing hazard patterns, a condition that makes urban planning for risk reduction and adaptation even more challenging (Box 3.1, no. 6).

In sum, the urban fabric can negatively influence hazard occurrence by intensifying existing hazards, creating new ones, increasing hazard exposure and compound hazards, constantly changing hazard patterns and fostering climate change, which reshapes hazard occurrence, thus magnifying global and local insecurity (see Box 3.1).

So let's have a look at the second risk factor (see Figure 3.1). Before continuing to read, ask yourself: how does the urban fabric – and related planning practices – influence location-specific vulnerabilities?

There are many different ways in which the urban fabric can increase location-specific vulnerabilities – and ultimately increase disaster occurrence. They are described in the following text and are summarized in Box 3.2.

First and foremost, many of the physical features that characterize the urban fabric can themselves be seen as vulnerability factors (Box 3.2, no. 1). If risk is not proactively reduced in due time, hazards are generally more disastrous if they occur in areas with high concentrations, closeness and interdependence of populations, political and economic centres, buildings, services and infrastructure (Figure 3.4). The simple explanation is that the more there is, the more can be destroyed. Furthermore, cities not only concentrate large numbers of people, they also tend to concentrate highly vulnerable people. These include the urban poor,[6] marginalized groups, and individuals weakened by conflict, malnutrition, and HIV/AIDS or other diseases, who often seek a better life in cities. To make matters worse, citizens are generally highly dependent on technical infrastructure and the services offered by political and economic centres. Consequently, the disruption of these (such as transportation and banking systems) can cause urban societies to collapse. Many (poor) city dwellers are further highly dependent on housing as a productive asset for pursuing their livelihoods. The destruction of housing thus also means a loss of income and livelihood, which makes them highly vulnerable. Moreover, aspects such as high densities, space restrictions, increased consumption, people's economic specialization as opposed to diversification and urban–rural linkages all contribute to a lifestyle that makes city dwellers highly dependent on others and, in the face of disasters, more vulnerable.

Urban–rural linkages refer to the two-way flow between urban and rural settlements consisting of people (e.g. moving or commuting between urban and rural settlements), money (e.g. financial remittances from the city to rural family members), goods (e.g. agricultural products flowing from rural-based producers to urban markets, and manufactured or imported products flowing from urban centres to rural settlements), infrastructure (e.g. shared roads), information (e.g. about prices and employment opportunities), and water and waste (e.g. shared river systems or urban waste being disposed of in rural areas). These spatial flows also coincide with flows between different urban- and rural-based sectors, such as agriculture, manufacturing and services – connections that are crucial for a vibrant local economy (Tacoli 2003). Against this background, even disasters that occur in rural areas may create serious implications for urban residents. A simple example is that of shared roads which, when impacted by disasters in rural areas, may leave cities 'cut off' from vital flows.

Nevertheless, it has to be noted that many of the urban fabric's physical features and related interdependencies can themselves also become key assets needed to reduce existing vulnerabilities – if they are mobilized for achieving risk reduction and adaptation (such as the number of people and the concentration of services, technical and social infrastructure, and political and economic centres).

Second, there is a set of different aspects that indirectly create susceptible conditions in cities as there are social, economic and political aspects that shape the urban fabric in such a way that it becomes vulnerable (see Box 3.2, no. 2). Examples are the use of inadequate construction materials and techniques and the unfavourable orientation of structures in space. This

Figure 3.4 Densely built houses and services in Caracas, Venezuela. Source: Ebba Brink.

can partially be attributed to high densities and space restrictions, which constrain the layout, design and location of the built fabric. Space restrictions can, for instance, make it difficult to adapt buildings to prominent wind or solar directions (e.g. to catch a breeze or avoid direct midday sun), and the available building sites might not allow deep foundations (to withstand landslides) or might just be too small for any mitigation works (for instance, for slope stabilizations or natural resource management). In addition, with urban areas being exposed to multiple hazards, disaster-resilient buildings are something relative and thus hard to achieve. Houses that are built to resist one specific hazard are, in fact, often vulnerable when facing other types of hazards. Earthquake-resistant buildings, for instance, require features in their constructions that are generally inadequate to withstand windstorms.

The use of inadequate construction materials and techniques can further be attributed to the following aspects:

- The unwillingness of both residents and politicians to improve dangerous living conditions and invest in risk reduction and adaptation measures due to (a) the false perception that cities are secure places, (b) the understanding that hazard-resistant design is too costly and (c) the strong reliance on, and confidence in, the existing institutional structures as well as the social security and welfare systems in place.

- Inability to invest in appropriate risk reduction and adaptation measures due to (a) people's limited access to information about resources and measures for adequate risk reduction, (b) slum dwellers' stigma and exclusion from the labour market that reduces their ability to finance their own measures and (c) lack of secure tenure, which can discourage the urban poor from investing in any long-term improvements.
- Importance of status, which can result in modern-looking houses which lack technical safety features.
- Corruption in the construction sector.
- The strong influence of the informal construction sector and related lack of quality control by city authorities and planners (see Section 4.3).
- The combined economic and spatial constraints that indisputably affect building qualities in many ways. Local livelihood practices are, for instance, not always suited to densely built areas with very limited space and can result in substandard modifications of the built environment to permit economic activities (such as the removal of supporting walls or the creation of landfills). (See Wamsler 2006c.)

Inadequate infrastructure for water supply, water and wastewater drainage, and waste management are further aspects that create susceptible conditions in cities and have been shaped by a multitude of social, economic and political factors. Lack of maintenance, high densities, space restrictions and fast urbanization processes are among the many causes. Urban water distribution systems often lose a significant percentage of water due to leakage from cracks, holes or corrugated pipes. This, together with the higher water consumption patterns that generally accompany urban lifestyles, adds to problems such as water scarcity and droughts. Meanwhile, inadequate disposal of waste, wastewater and rainwater can lead to blocked drains, eroded soil conditions, infiltrated walls and the creation of breeding grounds for diseases (e.g. by attracting vectors such as mosquitos and rats). Unsound management of urban waste, wastewater and rainwater may even result in negative impacts in rural areas (such as pollution, water scarcity, floods and destroyed harvests in downstream areas), with 'rebound effects' for cities (such as disruptions in the urban food supply). Increased risk and associated disasters are the outcome.

The third aspect listed in Box 3.2 is the increase of cities' vulnerability by creation of a domino effect that quickly spreads damage and secondary hazards, which means that even minor hazard events can trigger large-scale urban disasters. The characteristic features of the urban fabric as well as its interlinkages with environmental aspects create such a domino effect (see Section 3.1). High concentrations and increased densities and heights of the built fabric allow damage to be spread easily from building to building. Examples are fire spreading from one roof to the next or earthquake-affected houses falling on neighbouring buildings and infrastructure. Under such circumstances, secondary hazards (such as floods and fire) are also easily created (see Section 2.1). In addition, the quickly spreading damage combines with local conditions to instantly cause environmental hazards and further aggravate disaster impacts. Minor floods can transform into large-scale disasters when they occur close to waste disposal sites or leaking wastewater pipes, conditions that easily translate into the contamination of whole settlements.

The destruction of existing (natural) hazard protection is again another way in which the urban fabric increases cities' vulnerabilities (Box 3.2, no. 4). Examples are the demolition or reduction of natural windbreaks, flood walls, floodplains, slope stabilization, fresh air corridors and vegetation that is crucial for ground water generation and ground stability, permeability and cooling. The destruction of cities' natural hazard protection can also be contingent on urban–rural linkages, such as deforestation or destruction of wetlands in cities' surrounding areas (being, for instance, employed for the disposal of urban waste).

Yet another aspect of how the urban fabric can increase cities' vulnerability is its influence on socio-economic and environmental factors, which make urban citizens unable to prioritize and take measures to reduce or adapt to increasing risk (Box 3.2, no. 5). Whilst urban residents are, in contrast to rural dwellers, often less aware of their dependence on natural processes and the environmental impacts of their actions (McGranaham *et al.* 2004), they are more likely to live under conditions that make them unable to take on additional stress such as hazard or disaster impacts. The urban fabric's influence on noise, contamination, lighting, access to green areas and the quality of water and sanitation has, in fact, many negative impacts on people's health and well-being. Social segregation, inequality, urban violence and food insecurity present additional stressors (see Section 3.1 and Tables 3.1–3.2). Other issues that can make city dwellers unable to prioritize and take measures to reduce or adapt to increasing risk are the lack of local leadership structures and social cohesion, both of which can negatively affect any such joint efforts.[7] The urban fabric's characteristic features (such as architectural details or lack of public space and recreation areas) can contribute to this situation by creating either barriers to or 'boosters' for social cohesion.

On the bright side, the urban fabric also positively influences socio-economic aspects that can enable citizens to prioritize and take risk reduction and adaptation measures. The concentration of and access to technical infrastructure (for transportation, water and electricity), social services (such as education) and income opportunities (in form of wage labour) together with the density, flow and diversity of people, with their lifestyles and ideas, may be contributing factors to the often increased levels of gender equality, formal education and (formal) incomes in cities (see Table 3.3). The latter have all been linked to reduced risk and lower numbers of disaster fatalities (Wamsler *et al.* 2012; see Section 2.1).

Finally, the dynamic urbanization processes constantly reshape the urban fabric (such as buildings and extensions of settlements) and thus vulnerability patterns, which makes them hardly controllable (see Box 3.2, no. 6). This is in addition to the fact that city dwellers' differential vulnerability is generally more heterogeneous (compared with that of rural dwellers). This relates to differences in income sources, levels of income and education, habits and values, national, ethnic or religious backgrounds, household sizes and composition, housing types and access to services and information.

In sum, the urban fabric influences vulnerability by directly and indirectly creating susceptible conditions due to its characteristic physical features and their interdependencies, by causing a domino effect of quickly spreading damage and secondary hazards, by destroying natural hazard protection and creating stressors that reduce people's ability to adapt, as well as constantly reshaping existing patterns of vulnerability (see Box 3.2).

So what about the remaining risk factors? Before continuing to read, ask yourself: how does the urban fabric – and related planning practices – influence existing mechanisms and structures for response and recovery?

There are many different aspects related to how the urban fabric can negatively affect existing mechanisms and structures for response and recovery – and ultimately increase the number of disasters (Box 3.3). In general, the more people, buildings, services, infrastructure and economic and political centres that are affected, the more difficult it is for a society to react and bounce back to former (or better) conditions. This is related to the sheer amount of people in need, as well as the loss of manpower, services, infrastructure, economic resources and governance structures required to assist those in need (Box 3.3, nos. 1 and 2). This situation results in increased requirements for urban authorities to have functional, complex response and

BOX 3.2

How the urban fabric can exacerbate location-specific vulnerabilities – and ultimately increase disaster occurrence

1. Direct creation of susceptible conditions as many physical features of the urban fabric and their interdependencies can themselves be seen as vulnerability factors (e.g. high concentration, closeness and interdependence of populations and vulnerable groups, buildings, services, infrastructure, and economic and political centres).

2. Indirect creation of susceptible conditions as social, economic and political aspects shape the urban fabric in such a way that it becomes vulnerable. Examples include:

    • Use of inadequate construction materials and techniques
    • Unfavourable orientation of structures in space
    • Inadequate infrastructure for waste, wastewater and rainwater disposal.

3. Creation of a domino effect of quickly spreading damage and related secondary hazards (i.e. cascade failure created by the concentration and combination of all types of urban vulnerability factors).

4. Destruction of existing (natural) hazard protections.

5. Increased vulnerability due to the urban fabric's influence on socio-economic and environmental factors which makes urban dwellers unable to prioritize and take measures to reduce or adapt to increasing risk. Examples include:

    • Negative impacts on people's health and well-being
    • Creation of stressors such as urban violence and food insecurity
    • Reduced social interaction
    • Increased dependence on others (see no. 1 in this box).

6. Constant changes in vulnerability patterns due to dynamic urbanization processes.

recovery mechanisms and structures in place. On a more positive note, cities can provide many assets for disaster response and recovery – if they are not strongly impacted and can be mobilized. Nevertheless, even if only a small section of a city is affected, the urban fabric poses numerous challenges for disaster response and recovery, which are described below.

In urban areas, the access to, and transport, collection and housing of the affected population are often seriously hampered, obstructing emergency access, emergency supply, evacuation and resettlement (Box 3.3, no. 3). Causes relate to the characteristics of the building fabric, lack of space, remoteness and marginalization of settlements and related socio-economic aspects. An urban fabric with densely built settlements and multistorey buildings translates, if destroyed, into an enormous quantity of falling objects and rubble that block streets, make outside places unsafe and require major logistics for secure disposal. Densely built settlements often lack roads or pathways with adequate width for emergency access and evacuation, and also simply the extension of cities can hamper quick and easy access to remote or marginal settlements.

Under such circumstances, additional and specialized response and recovery functions for rubble clearance and public security might have to be provided (Box 3.3, no. 4). The increased need for assuring public security also relates to socio-economic aspects (such as gender issues and differences in income or in national, ethnic or religious backgrounds), which may require preventive measures for issues such as looting, unrest, or violence against women and children (West 2006).

The fifth way in which the urban fabric can negatively affect disaster response and recovery is the lack of (accessible and affordable) space, making the adequate housing of disaster-affected populations extremely challenging (Box 3.3, no. 5). Experience shows that emergency shelters, refugee camps and post-disaster resettlements do not work well if constructed far away from people's former homes and livelihoods, whilst other secure solutions are hardly feasible in cities (Baehring 2011; UNHABITAT 2003). Environmental and socio-economic characteristics of the urban fabric further worsen the situation, making temporary camps and settlements in urban areas problematic: free services provided in camps are, for instance, often a pull factor for the urban poor; double-registration of affected people in nearby camps is frequent; and conventional inner-camp solutions such as small-scale agriculture for food supply are hardly possible (for example, due to space restrictions and poor inner-city soil quality).

Furthermore, cities concentrate many people who are not able to actively take part in any response or recovery efforts (Box 3.3, no. 6). This does not relate only to individuals who are weakened by, for instance, diseases, conflict or malnutrition. The urban fabric also hampers citizens' response and recovery efforts via its influence on environmental aspects such as noise, pollution, reduced natural lighting, low air quality, inadequate water and sanitation, reduced water quality and lack of green space, which have many negative impacts on people's health and well-being. Urban dwellers may feel stressed (or are sick) even under normal circumstances, making them unable to handle additional stressors. Some environmental aspects have a further, more direct influence on people's capacity to respond and recover (Box 3.3, no. 7). High noise levels can, for instance, impair people's ability to hear disaster warnings and reduced natural lighting, together with increased fog and cloud, can reduce visibility, for instance during electric power outages.

The eighth way in which the urban fabric can impact disaster response and recovery is related to the increase in some vulnerability factors that also have a negative influence on disaster response and recovery (Box 3.3, no. 8). This is mainly due to the influence of the urban fabric on socio-economic factors (see Section 3.1 and Tables 3.3–3.4) that result, for instance, in their inhabitants not receiving disaster warnings or not being willing to use official emergency shelter. The increase in nuclear family structures and anonymity, lack of social cohesion, little sense of community and social inequality hamper mutual help to respond and recover, and further increase public expectations and reliance on institutional assistance as well as social security and welfare systems. An example that illustrates the interconnection between these factors is the extinction of the *hasshar* tradition in Pakistan. Traditionally, it allows people to access emergency help from other community members in the form of shelter and labour – in exchange for a symbolic meal. With the increase of wage labour, changes in family structures and loss of social cohesion, people increasingly (prefer or feel obliged to) hire manpower instead of calling *hasshar* because it creates reciprocal obligations (CWSPA 2005). Another factor that can negatively influence disaster response and recovery in cities is the lack of local leadership structures, together with the existence of quite heterogeneous groups, making it hard to establish the needs, contacts and required logistics to adequately assist affected communities (Baehring 2011). Moreover, urban dwellers' increased economic specialization (as opposed to diversification) makes recovery for people who have lost their income source especially challenging, leaving them without any (temporary) alternatives.

Finally, the constant changes to the extent, composition and layout of the urban fabric make it difficult to access and maintain up-to-date information required for adequate response and recovery, such as databases and maps (see Box 3.3, no. 9 and Section 4.3).

In sum, the urban fabric requires highly functioning and complex mechanisms and structures for response and recovery as well as specialized functions in order to deal with the many urban challenges that it creates.

BOX 3.3

How the urban fabric can negatively affect existing mechanisms and structures for response and recovery – and ultimately increase disaster occurrence

1. Increased requirements for having highly functional and complex response and recovery mechanisms and structures due to the huge populations cities need to provide for and who, moreover, live in vast and multifaceted urban settings.

2. Increased susceptibility of the existing response and recovery mechanisms themselves as urban disasters also affect their centralized operations (i.e. required manpower, services, infrastructure, economic resources and governance structures).

3. Access to, and transport, collection and housing of the affected population are seriously hampered, which obstructs emergency access, emergency supplies, evacuation and resettlement (e.g. due to inadequate and/or damaged urban fabric, lack of space, remoteness and marginalization of settlements as well as related socio-economic aspects).

4. Need for additional and specialized response and recovery functions, such as rubble clearance and provision of security services for citizens.

5. Lack of (accessible and affordable) space, together with socio-economic aspects of cities, makes the adequate housing of disaster-affected populations especially challenging.

6. Concentrations of people who are not able to actively take part in any response or recovery efforts (mainly due to the urban fabric's influence on environmental factors that have negative impact on people's health and well-being). Examples include:

    *   Noise and other urban stressors: people are already stressed before the disaster and cannot handle additional stress factors
    *   Contamination, reduced lighting, inadequate water and sanitation and lack of green space causing different illnesses.

7. Increase of factors that reduce the capacity of people and institutions to respond and recover (due to the urban fabric's influence on environmental factors). Examples include:

    *   Reduction of natural lighting, fog and cloud: reduced visibility when no electric light is available
    *   Noise: people can miss warnings.

> 8. Increase of vulnerability factors that have a negative effect on response and recovery (mainly due to the urban fabric's influence on socio-economic factors), such as the reduction of interactions with neighbours, exclusion and segregation, little sense of community, lack of local leadership structures and economic specialization.
>
> 9. Constant changes to the extent, composition and layout of the urban fabric, which make it difficult to obtain up-to-date information required for adequate response and recovery (such as databases and maps).

These challenges include high population densities and space restrictions that hamper access to, and transport, collection and housing of the affected people, the sheer quantity of rubble to be removed, urban dwellers' often low response and recovery capacities, and constantly changing conditions which make it difficult to obtain required up-to-date information and maps (see Box 3.3).

All in all, the presented analyses show that many of the characteristic features of the urban fabric with its physical and related socio-cultural, economic and political/institutional aspects can strongly influence existing hazards, vulnerabilities and the mechanisms and structures for response and recovery. They are determinants for urban risk and associated disaster impacts, and thus need to be addressed when planning for sustainable urban risk reduction and adaptation.

The question of how the urban fabric and related planning practice can increase risk and disaster occurrence will be further discussed in Section 3.3 and, especially, Section 4.3 entitled 'Differences in institutional approaches: current planning practice and increasing risk'.

3.3 Disaster impacts on cities

The focus of this section is not on cities' influence on disasters but the reverse interrelation, i.e. on the effects of disasters on cities. More specifically, it explores how disasters and other climate change impacts affect the urban fabric and related planning practice.[8] Before continuing to read, ask yourself: what kinds of urban sectors are sensitive to disasters and climate change? How do climate-related and non-climate-related disasters impact the characteristic features of the urban fabric including its physical and related environmental, socio-cultural, economic and political/institutional factors (see Section 3.1)? And how do these impacts influence, in turn, cities' level of disaster risk?

All types of urban sectors are exposed and sensitive to disasters and climate change, including housing, recreation and cultural heritage, water and sanitation, energy, transportation, telecommunications, environment and natural resource management, waste management and social and public services (including education, healthcare, police and emergency services). Table 3.5 illustrates how these urban sectors can be impacted by disasters and climate change.

Many disaster and other climate change impacts have direct and visible effects on the urban fabric, ranging from damage to individual houses to the destruction of whole cities, being wiped out because of issues such as earthquakes, landslides or the rise in sea levels. Potential impacts on the characteristic features of the urban fabric include:

- Damage or destruction of building stock (including the reduction in residential and commercial buildings, cultural heritage, etc.)
- Depreciation and wear of construction materials
- Damage or loss of land (including real estate, parks, etc.)
- Modification of topographic aspects (such as river courses, sloping terrain, etc.)
- Malfunctioning or complete destruction of technical and social infrastructure (including infrastructure for water supply, sanitation, energy, transport, communication, education and health services)
- Increase in population densities and overpopulation due to destroyed houses and settlements (see Tables 3.1 and 3.5).

The listed impacts on the urban fabric are closely interlinked with environmental, socio-cultural, economic and political/institutional aspects.

Examples that show the close interrelation between disaster impacts on the urban fabric and the urban ecosystem include:

- Damage to the urban fabric leading to aggravated environmental degradation, such as the destruction of wastewater pipes, thereby contaminating groundwater, or damage to factories and transport vessels leading to the dispersion of hazardous chemical substances
- Damages to the integrity of urban ecosystems, which affect the urban fabric and in turn foster existing hazards. A concrete example is the

Table 3.5 Disaster and climate change impacts on urban sectors that lead to increasing risk

Urban sector	Examples of disaster and climate change impacts – leading to increasing urban risk
Housing, recreation and cultural heritage	Destruction of housing, the type of infrastructure that is most affected by disasters (Jacobs and Williams 2011)
	Damage or loss of parks and other recreation sites (e.g. playgrounds situated in low-lying, flood-prone areas)
	Damage or loss of heritage sites having negative effects on people's identity, tourism and, thus, cities' economic development
	Damage or loss of land and real estate due to sea-level rise, landslides, erosion, melting permafrost, etc.
	Modification of the landscape through, for instance, the change of the course of rivers or terrain inclinations after earthquakes or landslides
	Increase in population densities to take care of (e.g. due to environmental refugees) and overpopulation due to destroyed housing stock
	Acceleration of depreciation and wear of construction materials (e.g. stone and metal) due to, for instance, increased climate variability
Water and sanitation	Damage or destruction of infrastructure for water and sanitation
	Negative impacts on urban water availability and quality (e.g. due to sea-water ingress, related salinization, and shifts in demand and supply due to climate change), also aggravating conflicts between end users (general public, private sector, etc.)
	Contamination of drinking water wells (see also former aspect)
	Overburdened wastewater and sanitation systems
Energy	Destruction of power lines and other energy infrastructure
	Reduced efficiency of cooling for thermal power due to higher air and water temperatures and/or lower water levels (Mideksa and Kalbekken 2010)
	Changes in wind power and hydropower potential, e.g. due to changed precipitation and wind patterns
	Increased and/or changing demand for heating and cooling, leading to increased energy use and possible blackouts
	Electricity failure and power outages, also disrupting other vital urban flows and services such as healthcare, transport and water supply

Table 3.5 (*continued*)

Urban sector	Examples of disaster and climate change impacts – leading to increasing urban risk
Transportation and telecommunications	Destruction of roads, railways, bridges, pipelines, ports, data sensors and telecommunication networks Blocked access to houses or settlements Poor road/transport conditions due to e.g. heat, ice, winds, visibility or disaster impacts and, thus, increased accidents
Environment and natural resource management (including waste management)	Impacts on urban ecosystem, which are directly linked to human well-being, such as temperatures, air quality, humidity, vegetation growth (including pollen levels) and presence of disease vectors (e.g. mosquitoes) Due to a warmer climate, increased impacts of droughts and new pests on the urban vegetation (Tubby and Webber 2010), including impacts on natural disaster protections such as permeable surfaces and wetlands
Social and public services	Damaged or destroyed healthcare units and schools Impacts on the provision of health and social care, education, police, fire fighting and emergency services (due to changes in risk patterns and direct disaster impacts on service provision; see Section 3.2)

destruction of or damage to urban vegetation, which can lead to citizens' reduced access to recreation sites and reduced air quality as well as increased risk of landslides and heat stress.

Examples that show the close linkages between disaster impacts on the urban fabric and urban society and culture include:

- Destruction of school buildings, housing and personal belongings that impact the level of formal education of children (and, over time, urban societies as a whole) (Box 3.4)
- Aggravation of social stresses and shocks such as disease and psychological shocks, which affect urban developments at all levels. Concrete examples are community distress, family disruptions and burglaries after disasters due to damaged houses or increased overpopulation, or illnesses caused by wastewater entering houses (see also following point).
- Increased vector- and water-related diseases and reduced food security caused by destroyed or inadequate infrastructure (including streets, water and energy supply)
- Forced eviction of disaster-affected slum dwellers (with city or national authorities using disaster impacts as a pretext for slum eradication) (Box 3.4)
- Destruction of architectural heritage, which undermines people's collective quality of life and identity
- Increased number of climate refugees (including city-to-city migration).

Disaster impacts that show the close linkages between the urban fabric, the urban economy and urban governance structures are:

- Disruption of local and household economies, affecting people's investments in improving their living conditions (e.g. their housing and settlement) and pushing already vulnerable groups further into poverty
- Destruction of productive assets such as home-based workshops
- Governance problems at different levels, resulting in aid budgets being skewed towards the recovery of one group or sector as opposed to another, resulting in increased urban inequalities
- Aggravation of political stresses, leading to increased corruption, bureaucracy, political conflicts and rivalry, which affect construction quality and urban development at all levels.

The described disaster impacts on cities can strongly hamper sustainable development and result in many challenges for urban planning practice. In

BOX 3.4

Disaster impacts on the urban fabric and their socio-economic implications for informal communities in Rio de Janeiro, Brazil

Rogerio and his family have been living in a low-lying flood-ridden area of Rocinha, an informal settlement in Rio de Janeiro. When Rogerio studied in third grade, severe flooding damaged their house and buried most of their possessions, including ID cards and other personal documents needed to take part in Brazilian civic society. To help his family recover, Rogerio was forced to quit school and start working. Eventually, the family was able to move to another area of Rocinha where floods are less frequent, but Rogerio never managed to go back to school and has today a notably lower level of education compared with his peers.

In the same settlement, in the higher altitude area of Laboriaux, a landslide in 2010 buried a row of houses and killed two women. Soon the prefecture proclaimed that, due to the risk, all of Laboriaux was to be removed. The local school, where 307 children studied, was closed down and the authorities offered to buy the 823 houses in Laboriaux for token sums (the 'sold' houses were then demolished to prevent new people from moving in). This approach created tension between those who agreed to sell and the majority who did not want to move. Lourdes was one of the latter: 'Nothing has ever happened here before,' she said. 'The money they offer does not correspond to the real value of our house, nor the work that we have put into our community during the last 15 years. It will never buy us a house of the same standard.' People further feared being forced to move to another part of Rio, far from their jobs and social networks with new and hostile criminal gangs, which would all considerably increase their level of insecurity.

By Ebba Brink

fact, after disasters, city authorities and other urban actors often find themselves confronted with the following situations:

- Increased pressure for land and housing provision
- Increased number of people and settlements that demand and depend on city authorities' assistance (e.g. to access adequate rental housing, house or land ownership as well as for house and infrastructure maintenance)

due to erosion of livelihoods, savings and physical capital at the household level

- Modification of cities' landscape, which affects future planning (e.g. infrastructure developments)
- Increased need of resources for specific (already planned) urban developments because of disaster impacts such as environmental contamination (polluted soil, wells, etc.)
- Impossibility to continue with certain urban development programmes, such as land legalization processes due to increased and unacceptable risk levels
- Reduced capacity or functioning of urban institutions that are directly or indirectly affected by disasters (e.g. due to damaged office buildings, disaster-affected project measures, staff death, injuries or leave, or damaged institutional reputation)
- Death, (temporary) disablement or migration of key persons (and workforce in general) at the national, municipal, community and household level, leading to an erosion of social capital for urban planning and governance at all levels
- Disruption of national economies and related governance functions of planning authorities due to post-disaster expenses and relocation of development investments (e.g. investments aiming to provide sustainable access to safe housing, drinking water and sanitation being disbursed on emergency issues)
- Impacts on national fiscal and monetary performance, indebtedness, the distribution of income, and scale and incidence of poverty, all of which negatively influence the provision and financing of housing and infrastructure at all levels
- Lower output from damaged or destroyed public assets and infrastructure, resulting in fewer resources that can be reinvested in the built environment
- Increased temporary investment and hence activity in the (formal and informal) construction industry due to rehabilitation and reconstruction efforts, providing increased challenges for construction control
- Need to build on response and recovery efforts for sustainable development, for instance temporary housing and settlements for disaster victims that over time need to be transformed into, or replaced by, permanent solutions. (See also Table 4.10 on general challenges faced by city authorities and planners.)

From the analyses presented in this section it follows that disasters and climate change strongly impact cities' level of risk. Box 3.5 shows how the different

BOX 3.5

How disasters can negatively affect the urban fabric and related planning mechanisms – in relation to urban risk factors

1. Influence on existing hazards. Changed hazard patterns due to:

    *   Existing hazards intensified (e.g. increased soil instability after earthquakes)
    *   New hazards caused (e.g. modification of the landscape which influences issues such as closeness to the sea, rivers or declivities)
    *   Colonization of new hazard-prone areas (e.g. resettlement of affected populations from earthquake- to flood-prone areas).

2. Influence on existing location-specific vulnerability. Examples include:

    *   Increase in the number of vulnerable people to house and serve (e.g. due to increased pressure for housing and land, and the erosion of people's livelihoods, savings and physical capital)
    *   Increased population densities and overpopulation
    *   Reduced quality of urban fabric (e.g. of housing stock, infrastructure and construction materials)
    *   Increased poverty and urban inequality
    *   Loss of urban assets and city functions that cannot be recovered (e.g. cultural heritage and land being washed away).

3. Influence on existing response and recovery mechanisms. Examples include:

    *   Increased number of citizens who are not able to actively take part in any response or recovery efforts (e.g. long-term impacts on health)
    *   Loss of urban fabric and land needed for response and recovery (e.g. reduced space for evacuation and resettlement)
    *   Temporary malfunctioning or interruption of urban fabric and city functions needed for response and recovery (e.g. street network and centralized governance structures).

4. Influence on all risk factors, i.e. planning for hazard and vulnerability reduction and preparedness. Examples include:

    *   Reduced financial resources for urban planning, risk reduction and adaptation (e.g. development budgets that are diverted to response and recovery efforts)

> - Reduced institutional capacity for urban planning, risk reduction and adaptation (e.g. due to reduced manpower, damaged infrastructure and lost reputation of urban authorities)
> - Impossibility to continue certain urban development programmes (e.g. land legalization processes due to increased and unacceptable risk levels).

impacts negatively influence all four risk factors (see Figure 3.1 and Section 2.1). After disaster occurrence, cities are thus generally at heightened risk. Cities that are not disaster resilient are prone to long-term suffering after disasters and are then more prone to adverse impacts from even smaller hazard events. Fast and efficient response and recovery are crucial. In contrast, resilient cities can be in an even stronger position post-event. An example is the city of Chengdu, China, which was struck by a strong earthquake in May 2008. The devastating losses spurred the city to mobilize forces for a durable reconstruction, and in 2011 Chengdu was appointed by UNISDR as a 'Role Model City' for resilient development in the 'Making Cities Resilient' Campaign.[9]

3.4 Climate change: reinforcing the city–disasters nexus

As described in Sections 3.2–3.3, the intricate linkages between cities on one hand and disasters on the other cannot be described by a one-directional cause-and-effect relationship. The city–disasters nexus is a bidirectional relationship, which further shapes and is shaped by other processes (such as poverty or climate change). This can result in both 'virtuous circles' (having favourable results) and 'vicious circles' (having unfavourable results). Such complex interplays can be described in so-called feedback loops. Before continuing to read, ask yourself: how does climate change influence the interconnection between cities and disasters (Figure 3.5)? And how could you best illustrate this interconnection?

The close interconnection between cities and disasters is strongly reinforced by climate change, resulting in vicious reinforcing feedback loops of increasing urban risk and disasters. The various feedback loops are created by (a) the influence of climate change on disaster occurrence due to an increase in climatic extremes and variability; (b) the influence of disaster occurrence on climate change; (c) the influence of (inadequate) urban development on climate change; and (d) the influence of climate change on (inadequate)

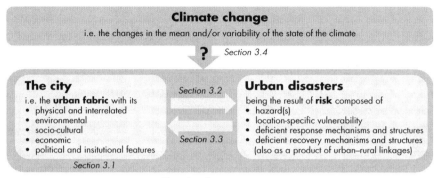

The city–disasters nexus

Figure 3.5 The city–disasters nexus: overview of issues addressed in Sections 3.1–3.4.

urban development due to its impacts on all risk factors (namely hazards, location-specific vulnerability and deficiencies in mechanisms and structures for disaster response and recovery) (Figure 3.6). Before continuing to read, have a look at Figure 3.6 and find examples for the illustrated connections.

The influence of climate change on disaster occurrence due to an increase in climatic extremes and variability. The most obvious link between climate change and disasters is the fact that climate change increases climate-related hazards, which includes climatic extremes and variability (see Box 1.2 and Section 2.1). Although not all disasters are directly attributable to climate change and increased greenhouse gas emissions, climate-related disasters represent on average two-thirds of all disasters,[10] and account for almost all the growth in the number of natural disasters since 1950 (Satterthwaite *et al.* 2007b). To make matters worse, the urban areas already at risk from disasters are those most likely to be affected by climate-related hazards in the future (IPCC 2007a; Moser and Satterthwaite 2008). Table 2.2 provides an overview of climate-related hazards in urban contexts.

Influence of disaster occurrence on climate change. Disasters influence climate change in that they can increase greenhouse gas emissions and reduce global warming (Figure 3.6, arrow connecting disaster occurrence and climate change). Examples are (a) wildfires and volcanic eruptions which release carbon emissions that were previously stored in biomass; (b) the destruction of forests or other land-use changes which reduces the availability of carbon sinks; and (c) volcanic dust and gases such as sulphur dioxide which are caused during major explosive eruptions and can result in a reduction of direct solar radiation and thus global cooling (USGS 2012).[11]

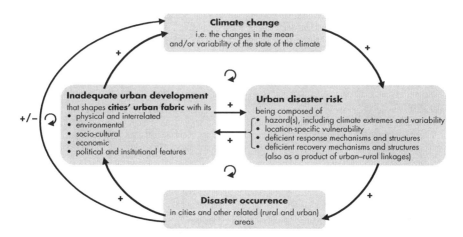

Figure 3.6 Causal loop diagram showing the simplified link between climate change, disasters and urban development. A causal relation between two variables A and B is portrayed by an arrow from A to B with a plus (+) or minus (–) sign. The sign indicates the type of change that occurs if variable A increases. A plus sign means that an increase in A causes an increase in B; a minus sign indicates that an increase in A results in a decrease in B. Reinforcing feedback loops are circular relations between various variables, all of which are connected with arrows going in the same direction. Reinforcing feedback loops are generally highlighted with bold arrows and indicate situations where vicious circles may arise.

Influence of (inadequate) urban development on climate change. Cities are major contributors to national greenhouse gas emissions (World Bank 2010), owing to the production and consumption activities of their residents, enterprises and institutions. Many sources claim that cities are responsible for 75–80 per cent of global greenhouse gas emissions (e.g. Munich Re Group 2004; O'Meara 1999; Stern 2006; UNHABITAT 2011). Such statements understate, however, the contributions from rural areas, arising from rural agriculture, deforestation, businesses and households that are not located in cities (Satterthwaite 2008). Irrespective of the specific numbers, urbanization in general and inadequate urban development in particular have a strong impact on the carbon cycle through the emission of greenhouse gases, aerosols and solid wastes, and land-use changes such as the creation of impervious surfaces, the filling of wetlands and the fragmentation of ecosystems (UNHABITAT 2011). In contrast, well-planned city development and urban form have the potential to effectively mitigate climate change. More compact cities, which use cleaner energy and are less dependent on motorized transport, have higher energy efficiency and thus produce less greenhouse gases (see Box 3.6).[12]

Influence of climate change on (inadequate) urban development due to its impacts on all risk components. Climate change negatively impacts urban development in many different ways. Table 3.5 provides an overview of climate change impacts on urban sectors. The most obvious and visible impact is the destruction of the urban fabric (including housing, infrastructure and services) due to climatic extremes and resultant disasters (see Figure 3.6). Other climate change impacts are less apparent. These can, for instance, relate to the increase in climate variability. An illustrative example is the expected increase in temperature and rainfall in some hot-humid climates, which is likely to lead to increased dampness in dwellings (as a result of transfers of dampness from the ground below the house). The humidity inside houses can become unbearable, with considerable consequences for human well-being (Davoudi *et al.* 2009). Other examples are rising temperatures which thaw out the layer of permanently frozen soil below the land surface (permafrost) causing the ground to shrink, or rising sea levels which cause water tables to rise and undermine the foundations of buildings (Wamsler 2008c). The result is damages to structures such as railway tracks, highways and houses, as well as landslides and erosion with further consequences for the urban fabric. Scientists have further pointed out that the melting of permafrost, which now covers large areas of Siberia and central Alaska, is likely to release large quantities of carbon, thus contributing to global warming and the creation of yet another climate change-related feedback loop with destructive consequences for the urban fabric (Zimov *et al.* 2006).

Apart from the described direct impacts on urban development caused by climatic extremes and variability, there is yet another type of climate change impact. These affect urban development due to their negative influence on the present location-specific vulnerabilities as well as the existing response and recovery mechanisms and structures. Examples are shortages of food, water and energy supply which may exacerbate existing tensions and conflicts between various end-users, and the increases of infectious diseases and other impacts on people's well-being (through air quality, road safety, etc.). Related urban–rural linkages are manifold (see Sections 3.1–3.3). Another example is the creation of millions of environmental refugees as a result of climate-induced disasters in interrelated (rural and urban) areas, for instance, due to sea-level rise, expanding deserts or catastrophic climate-induced floods or landslides. Increased migration (and the associated loss of livelihoods, conflicts and social disruption) is also one of several climate change-related factors, which are believed to contribute to the spreading of HIV/AIDS (UNAIDS 2008; UNHABITAT 2010b).

Additional climate change-related factors are increased poverty and malnutrition, leaving people weakened and possibly forced to choose between buying medication or food, as well as erosion of human, social and institutional capital, which undermine the work on HIV/AIDS prevention and the access to healthcare and education (UNAIDS 2008). The described examples clearly show how climate change can affect existing vulnerability, response and recovery and, in turn, negatively influence urban development (Figure 3.6).

There are further examples which illustrate how response and recovery mechanisms and structures can be negatively influenced by climate change, resulting in increasing risk and additional challenges for urban development (Figure 3.6). Increasing amounts of damage from climatic extremes and the associated insurance claims may put pressure on insurance companies to significantly raise premiums or even deny insurance for at-risk people. Other examples are the loss of labour force, municipal staff and related institutional capacities due to disaster occurrence or other disaster-induced aspects (such as the spread of HIV/AIDS), which can severely hamper existing mechanisms for disaster response and recovery, and the arrival of migrants (triggered by climate change impacts in other areas), increasing the number of people to provide for during disaster response and recovery.

In sum, climate change and urbanization, planned and unplanned, are deeply intertwined and frequently have adverse effects on each other. Climate change reinforces the city–disasters nexus as a result of its role in the increase in climate-related hazards, as well as in aggravating vulnerability and undermining existing mechanisms and structures for disaster response and recovery. Related urban–rural linkages are numerous. The resulting vicious and reinforcing feedback loops are shown in Figure 3.6 in the form of a causal loop diagram (commonly used in systems analysis and sustainability science). Such diagrams allow the illustration of non-linear relationships and related feedback loops, thus providing a useful tool to analyse the key variables and the causal relations influencing the city–disasters nexus.

3.5 Differences in cities' risk patterns

Based on the understanding of disaster risk described in Section 2.1, the risk of two cities located in distinct geographical areas can be compared by analysing (a) each area's current and future exposure to climate- and non-climate-related hazards (see Box 3.1); (b) related location-specific vulnerabilities (see Box 3.2); and (c) the mechanisms and structures that are in place for responding to and recovering from current and future hazards and resultant disasters (see Box 3.3). On the basis of this understanding, ask yourself: why are cities in low- and middle-income nations often considered at heightened risk when compared with cities located in high-income nations?

Over three-quarters of the 100 largest cities, of which most are located in low- and middle-income nations, are exposed to at least one natural hazard – a trend that is expected to continue in the next few decades (UNISDR 2004). Most of these cities are situated in Asia and Latin America. Of the 20 urban agglomerations worldwide projected to have the greatest hazard exposure of assets in 2070, 14 are in developing countries in Asia (IPCC 2012a). In general, low- and middle-income nations represent 85 per cent of the population exposed to earthquakes, tropical cyclones, floods and drought (UNDP 2004).

The high hazard exposure of many cities in low- and middle-income nations can firstly be attributed to their geographical location (e.g. on earthquake belts or in areas prone to sea-level rise). As an example, of the 3,351 cities in low elevation coastal zones around the world, 64 per cent are in developing regions (UNHABITAT 2008). Second, in these regions population growth drives hazard exposure, which becomes visible in the high number of people living on insecure land, such as steep slopes, unstable soil or riverbanks (see Tables 2.2–2.3 and Box 3.1).

Many southern cities are characterized not only by high levels of hazard exposure but also by high levels of location-specific vulnerabilities. This relates to the fact that, in contrast to more developed countries, southern cities tend to have lower income per capita, less advanced technology, lower levels of schooling and training, reduced access to information, less developed infrastructure and less stable and effective institutions (Easterling *et al.* 2004; see Section 2.1 on adaptive capacity). People's vulnerable living conditions reflect the shortcomings of local governments in terms of providing resilient infrastructure and services and their unwillingness to work with low-income groups, especially those living in informal settlements. Many southern residents (particularly in the poorest least developed countries where large proportions of the urban population live in unplanned settlements) are thus forced to live in substandard housing and have to deal with inadequate water and waste management services, overcrowding and limited access to financial resources, information and education. In addition, common environmental problems and stressors (such as pollution or inadequate water and sanitation that may generate illnesses and other burdens) often hinder city dwellers in low- and middle-income nations from actively taking part in any efforts to reduce their risk (see Box 3.2).

To make matters worse, many southern cities can further be characterized by low levels of aggregated institutional capacity to respond to and recover from hazards and disasters. The limited institutional capacity in low- and middle-income countries means, for instance, that analysis of climate change impacts at the city scale have generally considered only flood risk and not yet assessed additional potential impacts (IPCC 2012a), inevitably resulting in shortcomings for risk reduction and adaptation planning (see Box 3.3).

The combination of relatively high levels of hazard exposure and vulnerability, together with low institutional capacities to respond and recover, translates into the fact that during recent decades more than 95 per cent of disaster deaths have occurred in low- and middle-income nations (IPCC 2012a). There are, however, dramatic differences between cities and there is an increasing number of cities in middle-income countries (e.g. Bogota in Colombia or Mexico City in Mexico) where risk-reducing capacities exist that can manage increases in hazard exposure. But in much of the developing world, and particularly in the poorest least developed countries, such capacities are greatly restricted; meanwhile population growth further increases exposure (IPCC 2012a).

In a context of climate change and urbanization, the described differences between southern and northern cities in terms of the risk they face are, however, reducing. There are different reasons for this. First, climate change and urbanization are increasingly undermining the effectiveness of institutional responses and risk-reducing mechanisms in high-income nations, which were designed to be applied in the event of more 'usual' and more 'predictable' hazards and associated impacts. Second, in southern cities the low level of institutional capacity for risk reduction and adaptation can, to an extent, be compensated for by the rich range of innovative local-level responses developed by at-risk people. This is because many southern cities have a long history of dealing with disasters and coping with the limitations in national and local governments; however, the same is not true for high-income nations where local coping strategies are comparatively poorly developed and often considered harmful in terms of interfering with governmental and other institutional adaptation responses. This situation will be further discussed in the following chapters, which assess city authorities' current approaches to risk reduction and adaptation (Chapter 4), as well as city dwellers' own approaches to reduce and adapt to urban risk (Chapter 5).

3.6 Summary – for action taking

How do we conceive of cities in an era of increasing disasters and climate change? Cities can be perceived both as *at risk* and *as risk*, i.e. as being themselves at risk and also as being the cause of risk. While cities are not 'by default' at heightened risk if compared with rural areas, the lack of emphasis on the urban context in development and risk research, policy and practice has surely translated into lost opportunities to protect cities and the failure to adequately address the risk they face. Cities provide many important assets to address increasing disasters and climate change – but only if these are not strongly impacted and can be mobilized for risk reduction and adaptation planning.

Subtle differences in the urban fabric and related planning processes have a significant effect on how strongly a city is exposed to hazards, including climatic extremes and variability, and how it can withstand, counteract or overcome these hazards. Disasters are not one-off events caused solely by natural hazards but are generated by complex and interacting development processes in which the urban fabric plays a major role. The physical features that characterize the urban fabric can increase risk by intensifying hazards or creating new ones, exacerbating vulnerabilities and negatively affecting existing mechanisms and structures for response and recovery (see Figure 3.7 and Boxes 3.1–3.3).

Urban planning for risk reduction and adaptation thus needs to include measures that address all types of risk factors and at the same time target not only the physical features of the urban fabric but also related non-physical aspects which can turn cities into risk hotspots (Figure 3.7). A sustainable transformation of this kind can only be achieved if risk reduction and adaptation are mainstreamed to become an inherent part of urban planning practice.

In order to mainstream risk reduction and adaptation into urban planning practice, in-depth knowledge on the city–disasters nexus is required, allowing urban actors to modify their work so as to act upon increasing risk and, ultimately, achieve more disaster-resilient cities. The required knowledge on the city–disasters nexus includes an understanding on how disasters and climate change affect cities and, most importantly, knowledge on how cities' characteristic urban features influence climate change, risk and disaster occurrence. In this context, both urban–rural differences and urban–rural linkages require thorough consideration and are a reminder that adaptation and risk reduction cannot be seen as isolated urban or rural agendas.

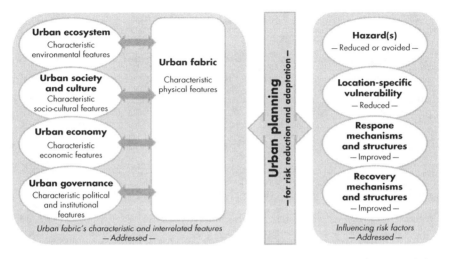

Figure 3.7 Urban planning for risk reduction and adaptation: reducing risk by addressing the urban fabric's characteristic and interrelated features.

Nevertheless, whilst it is evident that risk reduction and adaptation planning in cities needs to be seen through a particular urban 'lens' (to ensure that urban risk is adequately addressed), the combination of factors that determine urban risk and resilience is often too complex to assert with any degree of certainty whether a particular urban pattern is negative or positive to confront the effects of increasing disasters and climate change. A compact or a spread-out urban fabric does not, for instance, have any intrinsic value to reduce risk and adapt to (changing) climatic conditions. It is the local setting, the geographic location and its associated hazard conditions (e.g. high temperatures, high humidity, winds or floods) that determine what kind of urban fabric is the most appropriate. The planning paradox here is that an urban fabric that exacerbates climate change can, at the same time, adapt well to it (Box 3.6). And, the other way around, an urban fabric that mitigates climate change may not adapt well to increasing risk and disasters. In literature on sustainable development and climate change, the concepts of adaptation and mitigation appear, however, often as two sides of the same coin, making an implicit (but wrong) association between sustainable urban form and disaster-resilient cities (Davoudi *et al.* 2009).

The question as regards how disasters impact cities and, vice versa, how cities impact disaster occurrence, is also about what makes urban disasters unlike similar events in rural areas. In simple terms, urban disasters are unique because they occur in an environment that has adapted to absorb large populations and services, shaping cities' characteristic physical, environmental,

BOX 3.6

Sustainable urban form and disaster-resilient cities: adapting versus mitigating climate change

Mitigating climate change through the urban fabric is more easily achieved than adapting the urban fabric to it. Climate change mitigation is defined as taking actions to reduce greenhouse gas emissions to minimize climate change. In contrast, adaptation is aimed not at minimizing climate change but at taking action to reduce the effects of climate change (see Section 2.3). The compact city, which is advocated by literature on sustainable development, is the most appropriate urban form to reduce greenhouse gas emissions (Gonzales 2005). In contrast, the complexity of the interaction between the urban fabric, climate change and disasters (described in this chapter) rules out the possibility of a one-size-fits-all solution for adaptation. Furthermore, reducing risk associated, for instance, with extreme heat, cold, flooding and strong winds needs to be addressed by a mixture of urban design strategies that sometimes contradict each other. To make things worse, they may also not be the most appropriate for mitigating climate change (Davoudi *et al.* 2009).

Nevertheless, as the IPCC Fourth Report states, 'it is . . . no longer a question of whether to mitigate climate change or adapt to it. Both adaptation and mitigation are now essential in reducing the expected impacts of climate change on humans and their environment' (IPCC 2007a: 748). Sustainable urban development thus needs to be based on planning practice that incorporates both aspects (see Box 2.11). An example of how climate change mitigation and adaptation can be combined is the construction or repair of buildings to become both energy efficient and disaster-proofed. In addition, there are measures that can be classified as being part of both climate change mitigation and adaptation. Examples are green roofs and facades that can reduce energy consumption and CO_2 emissions, whilst at the same time minimizing rainwater run-off and temperatures and improving people's general and mental health. Such measures are called win–win measures.

Win–win, no-regret or low-regret measures can be defined as options that may produce positive results for several different fields or aspects. In the context of this book, win–win options are measures that, in addition to minimizing disaster risk or exploiting potential opportunities, also address other contributing risk factors and possibly have further social, economic or environmental benefits, including climate change mitigation. This also comprises those measures that are not primarily aimed at addressing disaster risk but

> nonetheless produce the desired adaptation benefits. In contrast, no-regret options are here measures whose socio-economic benefits exceed their costs regardless of how future climate change will manifest itself. Low-regret (or limited regret) options consist of measures of relatively low costs, which may have large benefits under projected future climate change (Hills and Bennet 2010).

socio-cultural, economic and political/institutional features. These features influence both disaster impacts and their management and relate to, amongst other things (a) cities' scale, (b) densities, (c) inhabitants' livelihood strategies, (d) economic systems and resource availability, (e) governance systems, (f) public expectations, (g) settlement structures and form, (h) likelihood for compound and complex disasters, and (i) potential for secondary impacts on surrounding areas (see Box 1.1).

3.7 Test yourself – or others

The following questions can be used to test yourself or others. The answers to most questions can be found explicitly in Chapter 3.

1. What makes a city different from rural areas? Name some of the distinctive physical, environmental, socio-cultural, economic and political/institutional features that characterize cities. In addition, highlight related urban–rural linkages.
2. What is an 'urban disaster' in conceptual terms?
3. Provide concrete examples of how cities' characteristic features relate to disaster risk. In other words: why can cities (or city dwellers) be at higher or lower risk compared with rural areas (or rural dwellers)?
4. How can urban planning practice influence (a) existing hazards, (b) location-specific vulnerabilities and (c) existing response and recovery mechanisms?
5. It is said that frequent disaster occurrence can cause a vicious circle of increasing risk. What does this mean? How can disasters increase citizens' level of disaster risk?
6. How can different urban sectors be impacted by disasters and climate change? Provide examples for three different sectors.
7. How can climate change influence disaster risk in cities? Provide concrete examples and explicitly relate them to the concept of risk.

8. What are the similarities and differences between planning for climate change adaptation and for climate change mitigation?
9. Identify an activity for urban development that can be seen as a measure for climate change adaptation and, at the same time, as a measure for climate change mitigation.
10. Could urban planning for risk reduction and adaptation undermine climate change mitigation? If yes, please give an example.

Case studies might also be a good basis for assessing the knowledge presented in this chapter.

Test yourself scenario 3.1: urban–rural differences in risk reduction and adaptation planning. You are working for an international NGO whose main expertise lies in risk reduction and adaptation in rural areas. The organization's board is planning to meet next month to discuss the organization's increased involvement in urban areas. In order to prepare for the meeting, your boss asks you to write a brief concept note on (a) the key differences between urban and rural settings, (b) how characteristic urban features (and their complex interrelations) can affect city dwellers' level of risk as well as institutional efforts to reduce and adapt to increasing risk, and (c) how characteristic urban features can or should be considered in the design and implementation of activities for creating more disaster-resistant urban communities.

Test yourself scenario 3.2: adaptation versus mitigation in urban areas. You are working for a local municipality and you are the focal point for integrating risk reduction into their urban development work. Your boss approaches you after having heard a radio broadcast titled 'The wrong way to a warmer world',[13] and asks you to clarify the difference and synergies between adaptation and mitigation and how this relates to your field of work.

3.8 Guide to further reading

The following references are a good starting point to follow up on some of the issues discussed in this chapter. Whilst the issues of cities, risk reduction and climate change adaptation enjoy a lively publication scene, which makes it generally hard to single out any specific book or article, hardly any publications address the combined domain of all three aspects (Box 3.7).

Benton-Short, L. and Short, J., 2008. *Cities and nature*. Abingdon and New York: Routledge.

Bosher, L., 2008. *Hazards and the built environment: attaining built-in resilience*. Abingdon and New York: Routledge.

Bulkeley, H., 2013. *Cities and climate change*. Routledge critical introductions to urbanism and the city. New York: Routledge.

IPCC (Intergovernmental Panel on Climate Change), 2014. Urban areas. In: *Climate change 2014: impacts, adaptation, and vulnerability*, Contribution of Working Group II to the IPCC Fifth Assessment Report (AR5) IPCC.

Pelling, M., 2003. *The vulnerability of cities: natural disasters and social resilience*. London: Earthscan.

Roberts, P., Ravetz, J., and George, C., 2009. *Environment and the city*. Abingdon and New York: Routledge.

Satterthwaite, D., *et al.*, eds., 2007a. Special issue on 'Reducing risks to cities from disasters and climate change'. *Environment and Urbanization*, 19 (1).

Shaw, R., and Sharma, A., eds., 2011. *Climate and disaster resilience in cities*. Community, environment and disaster risk management Vol. 6. Bingley, UK: Emerald Group Publishing Limited.

UNHABITAT (United Nations Human Settlements Programme) (2007). *Enhancing urban safety and security: global report on human settlements 2007*. London and Sterling, VA: Earthscan.

UNHABITAT (United Nations Human Settlements Programme) (2009). *Planning sustainable cities: global report on human settlements 2009*. London and Washington, DC: Earthscan.

UNHABITAT (United Nations Human Settlements Programme) (2011). *Cities and climate change: global report on human settlements 2011*. London and Washington, DC: Earthscan.

See also recommended readings included at the end of the other book chapters.

3.9 Web resources

Dynamics of Urban Change. This webpage provides urban practitioners with a collection of resources that emphasize the dynamic nature of urban change and the value and potentials of sharing knowledge that contribute to processes of economic growth, social development, cultural diversity, environmental sustainability and the reduction of poverty: http://www.ucl.ac.uk/dpu-projects/drivers_urb_change/home.htm

Global Facility for Disaster Reduction and Recovery (GFDRR). The GFDRR is a partnership of 41 countries and 8 international organizations committed to helping low- and middle-income countries reduce their vulnerability to natural hazards and adapt to climate change. Its Knowledge Center seeks to make disaster risk management information more accessible to practitioners: http://www.gfdrr.org

Knowledge Centre on Cities and Climate Change (K4C). K4C helps to keep track of what is happening in the field of cities and climate change by serving as a platform for sharing experiences and best practices as well as facilitating exchange of innovative initiatives: http://www.citiesandclimatechange.org

---

BOX 3.7

Books and journals on (urban) disaster risk reduction

The field of disaster risk reduction (DRR) enjoys a lively publication scene. Presently, there are close to 100 serial titles, which are all disseminated and read internationally. This tally includes journals that deal with natural hazards, disasters, emergency and business continuity management but, with a few exceptions, excludes the burgeoning field of climate change research. The scope, orientation, readership, level and style of the journals all vary markedly. Two-thirds of the journals focus explicitly on disaster risk topics. The rest provide a home for papers on such themes occasionally or frequently, but not predominantly.

Many of these serials are of a recent genesis and some were established only within the past two years. As many of these journals are in competition with one another, there is no available website to provide a guide to them. Barely a quarter of all journals are open access publications and the others require either payment or society membership to be read.

Although many of the journals deal sporadically with urban themes, few do so consistently. Among those that do, one can single out *Community, Environment and Disaster Risk Management* (thematic issues on urban topics) and the *International Journal of Disaster Resilience in the Built Environment*.

Geography, the earth sciences and sociology have dominated the production of scholarly monographs and textbooks in this field, but other disciplines have begun to make their mark. These include methodological ones such as spatial data representation (because geographical matters constitute a large part of hazard assessment and emergency management) and professional fields such as psychiatry and urban planning.

By David Alexander

UNHABITAT's Cities and Climate Change Initiative (CCCI). The United Nations Human Settlements Programme (UNHABITAT) is the UN agency for sustainable urban development. CCCI targets medium-sized cities in low- and middle-income countries and seeks to enhance the preparedness and mitigation activities of cities through good governance and practical local initiatives. Its Knowledge Centre helps to keep track of what is happening in the field of cities and climate change: http://www.unhabitat.org/categories.asp?catid=550

UNHABITAT's Disaster Management programme. This programme helps governments and local authorities to rebuild in countries recovering from war or natural disasters. It also provides technical assistance to help prevent future disasters and crises arising from natural hazards: http://www.unhabitat.org; http://www.unhabitat.org/categories.asp?catid=286

Urbanization and Global Environmental Change (UGEC). UGEC is a core project of the International Human Dimensions Programme on Global Environmental Change (IHDP) that is seeking to provide a better understanding of the interactions and feedbacks between global environmental change and urbanization: http://www.ugec.org

See also web resources included at the end of the other book chapters.

Part 2
Current practice

4 City authorities' approaches to urban risk reduction and adaptation

Learning objectives

- To gain a systematic overview of current practice in urban risk reduction and adaptation
- To assess city authorities' prevailing measures and strategies for addressing increasing disasters and risk, including climatic extremes and variability
- To explore how sustainable urban development practice that fosters risk reduction and adaptation can flow between different cities and communities (including high-, middle- and low-income, formal and informal communities)

The previous chapter has shown that the urban fabric has a significant effect on how strongly a city is exposed to different hazards (including climatic extremes and variability), and how it can withstand, counteract or overcome these hazards. Urban planning thus can, and should, play a major role in reducing and adapting to increasing risk and disasters. This chapter examines whether this understanding is in line with current practice. Before continuing to read, ask yourself: have many city authorities and planners seriously taken on their responsibility for enhancing risk reduction and adaptation? Do you know of any concrete actions taken?

The identification and analysis of concrete actions is very challenging as, even in countries where urban risk reduction and adaptation have been acknowledged as an important issue for urban planning, little has so far been done (e.g. Bicknell *et al.* 2009; Carmin *et al.* 2012; Greiving and Fleischhauer 2012; Mickwitz *et al.* 2009; UNISDR 2012c). National strategies for risk reduction and adaptation, if they exist, have in most cases not yet translated into planning practice. Independent of the specific top-down or bottom-up planning approach taken, integrated approaches hardly exist (UNISDR 2012c). Hence, while the important role of urban planning for risk reduction and adaptation is generally recognized, practices are still poorly developed. But how can we analyse in a systematic way what has been done so far? And how can we assess whether or not certain city authorities and planners have managed to tap into their full potential and learn from their experience?

On the basis of the theoretical understanding as regards risk, risk reduction, adaptation and mainstreaming presented in Chapter 2 and the city–disasters nexus presented in Chapter 3, planning for risk reduction and adaptation needs to address all contributing risk factors (i.e. hazards, location-specific vulnerability as well as deficiencies in existing response and recovery mechanisms and structures) and, in this context, consider the urban fabric's physical and related non-physical features. This can only be achieved in a sustainable way if risk reduction and adaptation become an inherent part of urban planning practice, which requires the use of a set of different mainstreaming strategies (Figure 4.1).

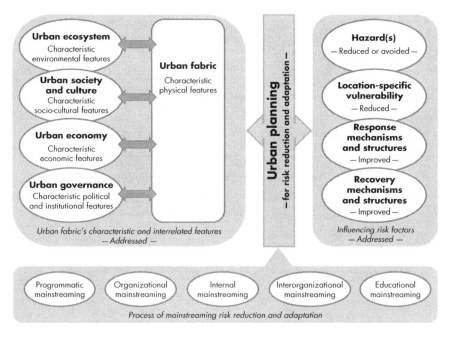

Figure 4.1 Achieving urban planning for risk reduction and adaptation – in theory.

However, the review of current approaches for urban risk reduction and adaptation presented in this chapter reveals a different picture. City authorities' current measures and strategies were identified by using a meta-evaluation of recent cross-country studies. This evaluation was complemented with information-rich case studies at country and city level.[1]

4.1 Overview of prevailing measures

The measures for risk reduction and adaptation taken by city authorities and planners were analysed as regards the different risk factors they address (see Figure 4.1 and Chapter 2) and how they consider the city–disasters nexus (see Figure 4.1 and Chapter 3). Before continuing to read, ask yourself: what are currently the most prevailing measures taken? And what are the key similarities and differences between the measures employed in cities in low-, middle- and high-income countries?

Despite huge variations between cities and countries, various patterns can be identified. They are described in the following text and are summarized in Table 4.6.[2] Note that differences between city authorities' measures for risk reduction and adaptation in low-, middle- and high-income countries will be further discussed in Section 4.3.

The first observation is that the hazard-specific measures taken by city authorities around the world are very similar and, thus, quite independent of the particular southern or northern context. Examples are measures such as the use of vegetation to reduce temperature, increasing the height of electricity installations in flood areas, marking former flood levels on houses for awareness raising, and rainwater harvesting and storage to cope with water scarcity or droughts. Tables 4.1–4.5 present an overview of the wide range of existing measures. Differences in these measures relate mainly to the urban areas' level of exposure, which, in turn, translates into distinct details and specifications (Table 4.6).

Second, most of the implemented or proposed measures are physical interventions, i.e. so-called 'physical', 'grey' or 'hard' measures that aim at reducing hazard exposure or at reducing the structural vulnerability of buildings and infrastructure to better withstand hazard impacts (see Tables 4.1, 4.2 and 4.6). City authorities take this grey infrastructure approach directly through the implementation of physical measures when building new settlements or public infrastructure, as well as indirectly through the revision of related legislation (such as building codes or tax incentives) or the elaboration of guidelines for the construction of disaster- and climate-resistant

Table 4.1 Measures taken by city authorities for hazard reduction and avoidance[a]

Hazard	Illustrative examples of activities
Precipitation, flood and sea-level rise (including related erosion and landslides)	Flood defences outside cities: flood walls, dams, dikes and building fronts to protect against flooding
	Green areas outside cities for flood retention, including the use of appropriate agricultural and forest practices
	Breakwaters to reduce erosion
	Reduction of the coastline
	Nourishment of beaches (putting sand off-shore or directly on beach)
	'Re-naturalization' of rivers and wetlands
	Draining of rising lakes to reduce the risk of outburst floods
	Establishment of water fascines/bundles to reduce erosion (helping to stabilize banks, trap sediment, promote revegetation, deflect water currents and slow sheet flow velocities)
Temperature, heat- and coldwaves	Reduction of heat island effect through vegetation, improved urban design and more adequate construction materials and colours (see also Table 4.2)
	Transport management to reduce air pollutants and related heat
	Reduction of heat emissions from residential and industrial buildings (e.g. by promoting more energy-efficient electrical devices and/or behaviours)
Drought and water scarcity	Measures to increase the recharge of ground water such as small dams or bunds, or underground dams that use recharge holes (more common in rural areas)
Wind- and snowstorms	Windbreaks outside cities: planning forestry areas upwind of towns
	During design stage of new urban developments, testing of the layout such as heights and shapes of buildings to avoid wind tunnel effects
	Controversial efforts such as cloud seeding/weather modifications to prevent hurricanes
Earthquakes and volcanic eruptions	(No hazard reduction for earthquakes or volcanic eruptions)

Table 4.1 (continued)

Hazard	Illustrative examples of activities
Fires	Firebreaks in urban design, such as buffer zones between heathlands and urban areas
Multi-hazard measures[b]	Keeping risk areas free from (further) developments through the enforcement of land-use plans
	Relocation of critical infrastructure such as schools and hospitals to (more) secure areas
	Control of the colonization of risk areas through local monitoring groups
	Signposting of risk areas to prevent their colonization
	Tax inducements to incentivize people to move to low-risk areas
	Transformation of high-risk areas into parks to prevent their colonization
	Land exchange, land pooling and land readjustment programmes to reduce citizens' hazard exposure
	Exchanging rights schemes to reduce citizens' hazard exposure
	Provision of information on hazard avoidance and reduction

Notes:

[a] The categorization of different measures is not always clear-cut and there are no doubt overlaps (see Tables 4.1–4.5).

[b] All types of measures for climate change mitigation can be considered to be part of hazard reduction; these have not, however, been specifically analysed for, or included in, this table which focuses on adaptation.

Table 4.2 Measures taken by city authorities for vulnerability reduction[a]

Hazard	Illustrative examples of activities
Precipitation, flood and sea-level rise (including related erosion and landslides)	Improved drainage systems (e.g. increased capacity and sewage systems with storage)
	Open water channels to funnel water flow and increase drainage and water storage capacity
	Retention ponds or other water collection points as temporary reservoirs to collect surplus water
	Creation of 'floodable zones' or 'water squares': low-lying spaces that can be used for temporary water storage or upstanding kerbs as flood retention space
	Creation of natural ecosystems buffers for vulnerable water bodies and low-lying areas inside the city (to avoid contamination)
	Encouraging businesses to 'sponsor' green areas (being able to use them for marketing, e.g. by planting their logo with coloured plants)
	Construction of flood-prone infrastructures on higher ground (e.g. raised communication ways, house constructions on poles, putting electricity installations higher up, contruction of higher bridges due to prospective floods or sea-level rise, etc.)
	Floating houses or man-made islands with floating houses (which adapt naturally to changing water levels)
	Separate treatment of rainwater, disconnected from sewage
	Buildings with unoccupied open spaces on the ground floor, which can be flooded
	Flood defences/barriers inside cities or urban communities (e.g. building fronts to protect against flooding or elevated entrances of buildings and metro stations)
	Increase of green areas inside the cities: parks and gardens, wetlands, water bodies and green roofs to reduce run-off water[b]
	Avoiding/reducing impervious surfaces to reduce water run-off[b]
	Distribution of plastic sheets to help channel rain- and wastewater (and to protect slopes)
	Organization of cleaning days to reduce the risk of waste clogging water channels
	Banning or charging higher taxes for the use of plastic bags and other non-environmental friendly packaging materials to reduce contamination and mitigate floods
	Fixing the covers of gully holes so they cannot float away during floods (to prevent people from falling in)
	For road safety, implementation of remedial measures for improved visibility under different weather conditions (e.g. cloudbursts)

Table 4.2 (continued)

Hazard	Illustrative examples of activities
Temperature, heat- and coldwaves	Road surfaces that can resist greater variations in high/low temperature
	Using construction materials and colours that can increase reflectivity of building facades and roofs
	Use of architectural design that optimizes natural ventilation in and between buildings
	Architectural details of buildings and design of urban fabric to provide shade (e.g. roof overhang, etc.)
	Blinds or insulation to keep inside cool; exterior shutters and blinds for maximum cooling
	Ensuring that fresh air from green areas outside the city can flow into the city (fresh air corridors)
	Building railway tracks equipped with insulation and cooling
	Construction of bridges with electric heating cables installed in the deck to minimize the use of de-icing salt in the winter
Drought and water scarcity	Ensuring sustainable watering and use of drought-resistant plants in public green areas, e.g. planting of trees or vetiver grass that have long roots
	Rainwater harvesting and storage (including the building of local water storage facilities, rainwater storage in wetlands and water bodies for later use, recommendation to new-build homes to build large underground tanks to collect rainwater for domestic use)
	Supply of water from more remote areas (pipelines)
	Desalination plants
	Water pricing as incentive to use water more efficiently
	Property tax incentives to encourage households to recycle wastewater or to store run-off rainwater for domestic use
	Measures to reduce the great share of water lost in urban water supply networks (e.g. by using information from flow meters, pressure gauges, etc.)
Wind- and snowstorms	Retrofitting existing building structures, including bracing and strapping of roofs, adding recommended fasteners, ties and reinforcements, and making entry doors and overhead doors more wind-resistant
	Windbreaks inside cities or urban communities, architectural or with plants, e.g. building fronts to protect against wind, or corridors of (locally native) trees to provide shade and windbreaks for the comfort of pedestrians
	Wind protection of pedestrian areas by building roofs, solid or porous screens and airtight transparent tunnels
	For road safety, implementation of remedial measures for improved visibility under different weather conditions (e.g. strong winds)

(continued)

Table 4.2 (continued)

Hazard	Illustrative examples of activities
Earthquakes and volcanic eruptions	Earthquake-proof construction of buildings and critical infrastructure (e.g. schools, hospitals)
	Installing base isolators underneath buildings, i.e. layers of steel and rubber, which reduce the amount of seismic waves that reach the building during an earthquake
	Flexible pipes that can move without breaking, e.g. for gas and water
	Layout of building fronts and streets to protect or guide mud or lava flows away from critical infrastructure, residential areas, etc.
	Increase of roof inclinations to avoid overload caused by volcanic ash and use of roofing/construction material which does not oxidize
Fire	Increased distance between buildings and use of fireproofed materials
Multi-hazard measures	Instituting incentives to remove unsafe buildings or to upgrade their level of safety
	Creation of open spaces and recreation sites to foster social interaction and regeneration
	Construction of subterranean electric wires
	Adapting building and planning codes for different hazards types (e.g. adapting building codes to include insulation and shadowing to cope with heatwaves, promote flood-resistant materials and include wind load requirements in building codes) and their enforcement
	Provision of information on vulnerability reduction
	Urban agriculture (e.g. to reduce impervious surfaces in the city, promote social cohesion and reduce food stress for the urban poor)

Notes:

[a] The categorization of different measures is not always clear-cut and there are no doubt overlaps (see Tables 4.1–4.5).

[b] Because this measure reduces the amount of water that must be dealt with, it could also be categorized as hazard reduction.

Table 4.3 Measures taken by city authorities for preparedness for response[a]

Hazard	Illustrative examples of activities
Precipitation, flood, sea-level rise	Protected rooms on top floors (as emergency shelters for floods)
	Elevated platforms above water level for emergency 'accommodation'
Temperature, heat- and coldwaves	Making it possible to widen access to public (air-conditioned/heated) buildings during heat- and coldwaves (also *ad hoc*)
	Making it possible to widen access to pools and parks during heatwaves (also *ad hoc*)
	Mapping of cool places and related public information so that people can go to these places during heatwaves
	Establishing logistics for the distribution of water bottles or the provision of other sources of drinking water to the public (e.g. opening of water hydrants; also *ad hoc*)
	Storage of grit and salt in order to be able to keep roads and pathways free from snow and ice during winter
Drought and water scarcity	Back-up systems for emergency water supply
	Measures to temporarily reduce water demand (e.g. incentives or restrictions to reduce the amount of water used during drought seasons, capacity building to save water and improve water use efficiency)
	Promoting low-tech methods to filter water if need be (e.g. sand tanks)
Wind- and snowstorm	Protected rooms in basements (as emergency shelter for hurricanes)
	Provision of wind-safe public buildings as community shelters in vulnerable settlements where people's homes offer no protection
Earthquakes and volcanic eruptions	Protected rooms on top floors (for tsunamis)
	Regular public earthquake simulations
Fire	Fire sprinkling systems in public buildings and regulations that prescribe the provision of fire extinguishers
	Water storage for dousing fire
	Regular public fire simulations
Multi-hazard measures	Early warning: developing weather warning systems including timely detection or prediction, alert and notification technology and communication of warning messages
	Guidance for behavioural changes for better preparedness (e.g. not storing valuables in basements in flood areas)

(continued)

Table 4.3 (*continued*)

Hazard	Illustrative examples of activities
	Distribution of information on adaptive behaviour during disasters (e.g. during heatwaves)
	Identification of urban risk indicators for developing local early warning systems (note: hardly done)
	Establishment of contingency, hazard action and/or emergency plans in order to be able to meet the immediate safety, security and health needs of potentially affected populations
	Evacuation planning (e.g. building or maintaining access and evacuation roads to and from vulnerable areas and buildings)
	Construction of escape routes to emergency shelters
	Construction of emergency shelters
	Back-up systems for vital urban flows, such as water, food and energy (e.g. the creation of multiple energy lines, the interconnection of municipal and regional water supply systems, and urban agriculture)
	Reduction of dependencies on external services for energy (e.g. using solar panels, building passive houses or interconnecting regional energy supply systems)
	Provision and/or distribution of energy generators for those who experience electrical power outrage

Note:
a The categorization of different measures is not always clear-cut and there are no doubt overlaps (see Tables 4.1–4.5).

Table 4.4 Measures taken by city authorities for preparedness for recovery[a]

Hazard	Illustrative examples of activities
Multi-hazard measures	Use of material for critical infrastructure that can easily (or cheaply) be recovered after hazard/disaster occurrence (e.g. use of building materials in flood areas that do not absorb water and can be restored solely by rinsing them, or construction elements that are designed to break in a way that allows them to be easily repaired)
	Construction of houses that can easily be rebuilt
	Recovery planning (including issues such as the coordination between different actors, identification of areas for resettlement and debris deposit)
	Preparations in order to be able to recycle debris (so that organic material like trees and shrubs can be recycled into compost, dirt can be used as landfill cover, crushed concrete and asphalt can be sold for use as hardcore for road constructions and other construction material such as metal and brick can be sold or used for landscaping applications)[b]
	Support of disaster insurance that promotes risk reduction and adaptation (instead of risk taking)
	Promotion of index-based weather insurance schemes (e.g. in flood or drought areas)
	Promotion of community emergency funds or other post-disaster aid funds
	Offer of advisory services for reconstruction
	Developing guidelines for safe recovery (e.g. related to damaged electrical wiring after an earthquake or venomous animals and spreading of disease after floods)
	Social protection for affected families (e.g. unemployment insurance, sick leave, etc.)

Notes:

[a] The categorization of different measures is not always clear-cut and there are no doubt overlaps (see Tables 4.1–4.5).

[b] For more resources on recycling of debris, see www.epa.gov/osw/conserve/imr/cdm/debris.htm

Table 4.5 Measures taken by city authorities for risk assessment and awareness raising

Hazard	Illustrative examples of activities
Multi-hazard measures	Mapping of existing hazards (partially taking into account climate change scenarios) Mapping of vulnerable areas – separately for different types of hazards Identification of vulnerable groups and their location as basis for targeted action (note: hardly done) Identification of people's adaptive capacities (including used and unused capacities) (note: hardly done) Informing the general public about existing risk (e.g. distribution of information material and organization of information forums) Different types of risk awareness measures/campaigns (e.g. using environmental traffic lights for publicly showing the risk level of different urban communities [Figure 4.4], marking former flood levels on houses or 'transforming' former disaster areas into outdoor museums)

housing and infrastructure. Some city authorities have further entered into dialogue with insurance companies to support insurance schemes that demand compliance with specific constructive precautions (e.g. Wamsler and Lawson 2011). In general, the focus is on physical measures to improve housing, water supply and sanitation whilst other urban sectors seem to receive less attention (e.g. cultural heritage, transportation and telecommunications, energy, social and public services). The strong emphasis on housing relates to the fact that buildings are the type of infrastructure that is most affected by disaster and, in the case of earthquakes, the main cause of disaster deaths (Jacobs and Williams 2011).

Examples of grey measures that are directly implemented or advocated through legislation or guidelines include floodwalls and dams, breakwaters to reduce erosion, the construction of flood-prone infrastructure on higher ground, the use of buildings as windbreaks, improved drainage systems, upstanding kerbs as a flood retention device, subterranean electric wires, architectural design that optimizes natural ventilation in buildings, blinds or insulation to keep the inside cool, construction material that increases the albedo effect (i.e. reflectivity) of building facades, roofs and streets, and road surfaces that resist higher variations in temperatures and precipitation (see Tables 4.1–4.2). A concrete example of this last is from the city government of Mombasa in Kenya, which has been working on assessing and improving bridges and roads since 2006 to ensure that every road can withstand the increased rainfall associated with climate variability and long-term change (Bicknell *et al.* 2009).

Some of the grey measures that are common in many disaster-prone cities in the south are rather new for northern cities. The modification of existing buildings to create unoccupied and floodable open spaces on the ground level is, for example, not current practice in Sweden, but is now increasingly discussed for climate change adaptation. Likewise, the use of exterior window shutters or blinds, which keep sunlight away from the windowpane and thus maximize the cooling effect, is also becoming more and more prominent in northern Europe (e.g. Jha 2006).

The above-mentioned guidelines for supporting the construction of disaster- and climate-resistant housing and infrastructure are often developed by national rather than municipal authorities. Concrete examples are guidelines published by the Federal Emergency Management Agency (FEMA) of the United States Department of Homeland Security, such as the 'Homeowner landslide guide', the brochure on 'Protecting your home from hurricane wind damage', or the 'Earthquake safety guide for homeowners'.[3]

Unfortunately, city authorities' dominant grey infrastructure approach often addresses the physical factors of the urban fabric in isolation, i.e. without giving sufficient consideration to related non-physical factors, which can lead to an increase in urban societies' level of risk (see Table 4.6). Exceptions are the measures that focus on the link between the urban fabric and the urban ecosystem. In recent years, city authorities have increasingly started to search for 'green' and 'soft' measures to complement and partly replace the still dominant 'hard' solutions, a trend that can primarily be observed in high-income nations.

The third identified pattern is that most urban adaptation measures aim to improve water management of cities to reduce the risk of floods, landslides, extreme temperatures, urban drought and the urban heat island effect. Typically, they directly address the link between the urban fabric and environmental factors (see Table 4.6). In order to assist city authorities in these efforts, several guidelines have recently been elaborated (e.g. Jha *et al.* 2012; Loftus *et al.* 2011; Shah and Ranghieri 2012). Grey measures to improve water management include the construction, improvement or maintenance of dikes, sewerage and drainage systems, open water channels and retention ponds (see Tables 4.1–4.2).

In southern cities, grey measures are often considered to be the best solution to improve water management, whilst northern cities are increasingly taking a 'green and blue infrastructure approach' (which is an example of the previously mentioned trend to complement or replace physical measures with other approaches) (see Table 4.6). In southern cities, grey measures to

Table 4.6 Similarities and differences between city authorities' measures for risk reduction and adaptation in low-, middle- and high-income nations

Context	Measures for risk reduction and adaptation
Similarities: not context specific	Risk reduction and adaptation measures that are promoted in formal areas are very similar; differences relate mainly to the areas' level of hazard exposure
	Strong focus on grey infrastructure approach for urban risk reduction and adaptation
	Emphasis on reducing or avoiding hazards and reducing physical vulnerabilities
	Physical, social, cultural, economic and political/institutional aspects tend to be addressed individually
	Many efforts focus on water management through combined grey, green and blue measures, which address the linkages between the urban fabric and environmental factors (an exception to the previous point)
	Strong interest in: (a) keeping disaster-prone areas free from further development and (b) developing guidelines for climate-resilient construction of housing and infrastructure
	Preparedness for response receives little attention; preparedness for recovery even less
	Measures for risk reduction and adaptation are seldom based on actual risk assessments (see Section 4.2)
Differences: specific to high-income countries	More emphasis on environment-oriented measures that address the link between the urban fabric and urban ecology
	Focus on the implementation of win–win or no-regret solutions (such as green infrastructure) (see Box 3.6)
	Special interest in combining measures for risk reduction and adaptation with climate change mitigation measures (to reduce greenhouse gas emissions)
	Better social protection systems; higher level of social security (income security and access to essential services, in particular health and education)

Table 4.6 (continued)

Context	Measures for risk reduction and adaptation
Differences: specific to low- and middle-income countries	Greater focus on basic and technical infrastructure (e.g. to improve rain, wastewater and waste management)
	Less emphasis on green and blue infrastructure (and the link between urban and environmental issues)
	More hazard-specific measures as opposed to multi-hazard, win–win and no-regret solutions (as specific risk is more present/evident) (see Box 3.6)
	Greater involvement of NGOs, CBOs and citizens, also leading to increased attention being given to low-tech measures, temporary solutions, locally based measures and participatory approaches to risk reduction and adaptation planning (see Section 4.3)
	More focus on a variety of response (and in part, recovery) preparedness measures
	More use of temporary solutions (as opposed to long-term measures/strategies)
	Risk financing is seldom adequately integrated into (social) housing finance mechanisms (i.e. governmental and non-governmental subsidies, micro-credits and family savings, mutual help or self-help mechanisms), especially in marginal settlements
	More pro-poor interventions that link risk and poverty reduction and focus on asset enhancement, empowerment, livelihood support and health (which address the link between the urban fabric and socio-economic issues) (see Section 4.3)

improve water management are sometimes combined with low-tech measures to compensate for financial constraints. Examples are the distribution
of plastic sheets (to help channel rain and wastewater and to protect slopes
at risk from landslides) or the organization of cleaning days (to reduce the
risk of waste clogging water channels). As regards the latter, some cities
have also taken legislative action by restricting the use of materials that
end up clogging sewages and canals. As an example in Ho Chi Minh City,
Vietnam and Dhaka, Bangladesh polythene bags have been banned (Bicknell
et al. 2009; Schinkel *et al.* 2011).

The 'green and blue infrastructure approach' to improve cities' water
management often goes hand in hand with the 're-naturalization' of ecosystems. It underlies the increasingly promoted ecosystem-based adaptation
(e.g. Roberts *et al.* 2012). The essence of this approach is to work *with*
nature, not against it, thus addressing the link between the urban fabric
and the urban ecosystem (see Section 3.1). Examples are planning of
residential and commercial areas to also include open space, the avoidance
of impervious surfaces to aid water absorption, the preservation of wetlands to defend against flooding, the use of vegetation on roofs and vertical
surfaces (e.g. green walls) to reduce water run-off and absorb heat, the
re-naturalization of rivers, and the general increase, restoration and maintenance of green areas and open space (e.g. as so-called floodable zones)
(see Tables 4.1–4.2). In order to support these last, city authorities also
revise their urban regulations. Examples are the 'Seattle Municipal Code',
in which a special chapter has been included to regulate the city's parks
(the 'Park Code')[4] and the Biotope Area Factor (BAF) of the city of
Berlin, which expresses the ratio of the area covered by vegetation to the
total land area of any urban development (Kazmierczak and Carter 2010).
Another way to support the general increase, restoration and maintenance
of green areas comes from Lima, Peru where city authorities have given
local enterprises the task of taking care of a number of green slopes, which
they can then use for their marketing purposes (e.g. by planting their logo
with coloured plants).[5]

What makes the use of open, green and blue spaces for urban risk reduction
and adaptation so attractive is that they also improve citizens' quality of life
and provide recreational areas and places that can foster social interaction.
Green roofs and the provision of green and recreational areas were, for
instance, an integral part of the 'Eco-City Augustenborg' project in Malmö,
Sweden, described in Test yourself scenario 4.3 at the end of this chapter.
Apart from the 'Eco-City Augustenborg' project, many other examples can
be found to illustrate how green and 'soft' structures are increasingly

favoured in the northern hemisphere to replace the still dominant 'hard' solutions (see Table 4.6). A case in point is New York City. To avoid untreated stormwater and sewage regularly flooding the streets, the city authorities decided not to focus on improving the sewerage system (which would cost US$6.8 billion), but instead to invest in green infrastructure (which costs US$5.3 billion) (UNISDR 2011b). Another example comes from the Netherlands, where cobble shores are increasingly replaced by the use of beach or shoreface nourishment (Stive 2012). The latter is carried out by placing sand off-shore, resulting in 'growing beaches' that can respond to alterations in climate change forecasting, as they grow according to sea-level rise. The increased use of shoreface nourishment is also a good example to illustrate the ongoing shift from a 'direct protection approach' to a 'feeder approach'. Sand nourishment to avoid flooding and erosion is also promoted in some southern cities, for instance in Lagos, Nigeria. A lack of financial resources and institutional capacities, however, often hampers the desired results (UNESCO 2012). The cities of Manizales in Colombia and Pune in India are other examples that actively use a range of different green and blue measures (Hardoy et al. 2011; UNISDR 2012b). Nevertheless, in the southern context, grey measures are still often perceived as being 'more advanced' and are thus the preferred option.

An increasingly significant aspect of urban water management, in both northern and southern cities, is dealing with water scarcity and drought. More and more cities are establishing urban drought programmes aimed at finding alternative sources of water to temporarily increase supply and at reducing water demand (e.g. through sustainable watering and switching to drought-resistant urban vegetation).

The fourth identified pattern is that most of the assessed cities place great emphasis on hazard avoidance measures – by prohibiting further development in urban areas that are already prone to disasters and climate change (and which may be even more endangered in the future) (see Table 4.6). Typically, northern countries pursue this goal through the enforcement of land-use plans and policies or in some cases, the relocation of critical infrastructure (such as schools and hospitals) to more secure areas. In Dorset, UK, for instance, an urban policy was implemented that requests a 400-metre buffer around the heathlands to reduce the fire risk from bordering settlements and prevent new colonization in risk areas (Kazmierczak and Carter 2010). In the southern context, a lack of enforcement capacity often translates into a wider variety and combination of measures to gain control over the colonization of risk areas. Examples include small-scale measures, such as local monitoring groups and signposting (Figure 4.2), as well as more far-reaching measures, such as tax incentives, transformation of high-risk areas into parks, exchanging rights schemes as well as land exchange, land pooling and land readjustment programmes (see also under Section 4.3).

The fifth pattern is that preparedness for response and, to an even greater extent, preparedness for recovery receive little attention from city authorities and planners (see Table 4.6). Exceptions are efforts to improve early warning systems. Early warning includes different elements: the monitoring of hazards (such as sea-level rise, precipitation, temperatures, wind, snow

Figure 4.2 Signposting in San Salvador: a small-scale measure to reduce citizens' hazard exposure. The text on the sign states that the use of the indicated public green area is strictly forbidden. Source: Christine Wamsler.

and ice, or the progression of drought), hazard forecasting, the communication of forecasts (and risk reduction measures) to vulnerable population groups, and possibly simulations. The identification of, and communication with, vulnerable groups is often the weakest link of the chain (see DFID n.d.). City authorities' level of involvement in early warning activities varies greatly. National meteorological services and the emergency authorities often take the main responsibility. There are, however, some cases where city authorities themselves have taken the lead. An example is the Botkyrka municipality in Sweden described in Box 4.1, which has been driving the establishment of both a municipal and national warning system for heat-waves.

A wider variety of small-scale, low-tech and participatory early warning systems can generally be found in the southern context, whilst more high-tech and data-intensive systems seem to be the dominant choice in northern cities (Box 4.2). Data availability is often higher in northern cities. However, the identification and collection of specific urban risk indicators is, in both contexts, in its early stages. Such indicators are needed to supplement disaster databases and develop adequate early warning systems for cities (Wamsler 2006c). As regards the communication of potential risk-reducing measures to the public, very few cities have developed guidelines to inform their inhabitants about how to respond and adapt individually. Botkyrka municipality in Sweden is an exception to this (Box 4.1). It is more common for the national government to provide citizens with some guidance. As an example, the 'Get Prepared' campaign of the Canadian government encourages people to prepare themselves so that they can cope on their own for at least the first 72 hours of an emergency, enabling city authorities and other stakeholders to focus on those most in need.[6] Other response preparedness measures taken by city authorities include setting up emergency plans for health and social care systems, mapping places where people can cool down during heatwaves, setting up the logistics for the public distribution of water bottles, and offering existing public buildings as community shelters in vulnerable settlements where people's homes offer no protection (see Table 4.3).

Examples of response preparedness through grey measures include the construction of community emergency shelters, protected rooms in basements (for hurricanes) or on top floors (for tsunamis), access roads to vulnerable areas, and escape routes to emergency shelters (see Table 4.3). In addition, back-up systems for vital urban flows (such as energy, water and food) are crucial, but are rarely developed in a coordinated way, being left to isolated actions of single actors. The creation of multiple energy supply lines is one example of such a back-up system.

BOX 4.1

Botkyrka: the only Swedish municipality that was prepared for heatwaves in 2010

Many urban areas are expected to increasingly experience heatwaves. In spite of Sweden's relatively cold climate, fatalities occur during hot periods and their number is bound to increase in the future as heatwaves become more common and more extreme. Death rates significantly rise in Sweden already when the average temperature remains over 22–23 degrees Celsius for more than two consecutive days. Those who are most susceptible to heat are the elderly, people who take certain medicines or have certain illnesses and those who live on the highest and lowest floors of multi-family apartment blocks. By increasing emergency response capacity for heatwaves in the municipalities, it is possible to prevent susceptible residents from dying or becoming seriously ill.

Botkyrka municipality, located not far from the capital Stockholm, was the only Swedish municipality that was prepared for the heatwaves in 2010. Staff from the city planning department and social services had cooperated with researchers and produced a range of future scenarios based on a changed climate. On this basis, the municipality gathered information to the following questions:

- How many people in the municipality are especially susceptible to heat?
- Where in the municipality (i.e. in which districts) do these people live?
- How many of these people are service users, i.e. are in regular contact with the municipality through their different services?
- Does the municipality have any locations that can be used temporarily by those suffering from the heat to cool off?
- How can instructions for use before, during and after a heatwave be formulated for health and social care personnel?
- How can the municipality deliver information to heat-susceptible individuals ahead of a heatwave and how should the information be worded?

As a result, when the 2010 heatwave hit, Botkyrka was the only municipality in the country to have identified the number of inhabitants who are especially vulnerable to heat, and to have ready-made maps showing the whereabouts of heat-susceptible individuals, geographical analysis of how outreach activities can be planned and evaluated, checklists for municipal personnel for use in outreach activities, and even information sheets on the municipal homepage and for distribution to the public, providing concrete tips for individuals on what to do during the heatwave. Today, heatwaves are taken into consideration when projecting official buildings, such as homes for the elderly, and the elaboration of an official map to show cool places that are accessible to the public during heatwaves is under way. Finally, the work of Botkyrka municipality has also spurred on a national project led by the Swedish Meteorological and Hydrological Institute (SMHI), which aims to have a national warning system for heatwaves by 2013.

By Ingrid Molander and Annika Carlsson-Kanyama

BOX 4.2

Early warning systems: low- and high-tech measures

In Europe, early warnings are spread using the Internet, mobile phones and other media. As an example, on http://www.meteoalarm.eu, information on weather warnings from different European countries is exchanged and accessible to the public. In Denmark, Germany, Hungary and the Netherlands, citizens can receive text message alerts on upcoming severe weather events via their mobile phones. The Environment Agency of England and Wales has gone even further and offers flood warnings by phone, SMS, email or fax to which more than one million people subscribe.

For the institutions responsible for the warnings, intelligent communication systems mean that they can, in turn, gain access to the information at an earlier stage. The importance of combining high- and low-tech measures can be illustrated with the case of Hungary and Germany where groups of volunteers, so-called spotter groups, are used to support the meteorologist services with accurate and real-time reports on local events (Rauhala and Schultz 2009).

In many low- and middle-income countries, there are long-term experiences with community-based local early warning systems. In the capital of El Salvador, such systems have long been established. The result is a direct and permanent communication (round the clock) between communities, municipality and the Centre for Hydrological Forecast and Emergency Services. Communities are responsible for measuring, monitoring and communicating water levels of rivers and rainfall, as well as for warning citizens with the help of sirens and door-to-door notice. This system allows communities to be informed about floods around 1 hour and 30 minutes ahead of their occurrence, which provides sufficient time to get prepared and avoid fatalities (RedDesastres n.d.; Wamsler and Umaña 2003).

Furthermore, cities can prepare for something as extreme as a complete system failure. This requires measures to reduce cities' dependency on external services for water, energy or food supply. Measures to reduce water and energy demands or to generate energy locally are, for instance, the construction of passive houses, the use of solar panels, the promotion of district heating and cooling systems, the improvement of water use efficiency, the construction of local water storage facilities, and the interconnection of municipal and regional water and energy supply systems (CAP 2007) (see Table 4.3). Urban agriculture can be seen as an adaptation measure aimed at reducing dependencies on food supply systems, whilst also reducing surface sealing and increasing social cohesion and quality of life. This is, for instance, promoted by Lund municipality in Sweden.[7] In the southern context, the support of urban agriculture for risk reduction and adaptation is closely linked to poverty reduction measures (see Tables 4.2, 4.3 and 4.6). In Belo Horizonte, Brazil, for instance, urban agriculture is part of the city authority's strategic multi-stakeholder plan aimed at contributing to the food and nutritional security of the city's poorest residents. Measures include the promotion and establishment of school and community gardens, a pro-orchard programme, workshops on planting in alternative spaces and training workshops on food and nutritional security, benefiting a total of around 147,000 citizens.[8]

Owing to the lack of adequate response preparedness, many city authorities act in an *ad hoc* fashion during disasters. An illustrative example is the heatwave of summer 2011 in New York, during which citizens suffered from high temperatures and problems in water supply. The city authority reacted by extending the opening hours of public pools and parks, converting fire hydrants into temporary drinking sources and advising inhabitants without air-conditioning to visit cooling centres (Kleinfield 2011). Other issues city authorities often have to solve on an *ad hoc* basis (because related preparedness planning has not been carried out) are different actors' roles and responsibilities, the financial management of external support (such as donations), the location and type of construction for temporary shelter and resettlements, and the disposal and management of debris.

As for recovery preparedness, hardly any concrete actions are taken. Many benefits of social protection and welfare systems (such as unemployment insurance and paid sick leave) can, however, be seen as recovery preparedness measures, although they are not primarily aimed at risk reduction and adaptation (see Box 5.2). They assist citizens in quickly bouncing back after disasters. Another, more direct, measure is the use of construction materials that can easily be recovered after disasters (for instance, in public buildings, power distribution infrastructure, roads, railways or pipelines) (Table 4.4). In addition, there is an ongoing discussion around the importance of creating (micro-)insurance schemes that support disaster recovery but do not foster risk taking, so-called 'moral hazard'.[9] In southern cities, a variety of related measures can be found, such as index-based weather insurance schemes or collective insurance policies. As an example, in Malawi, index-based weather insurance schemes are used in which the claims are independent of losses and based on an index of local rainfall (Mechler and Bayer 2006). In Manizales, Colombia, the city authority has established an insurance programme for buildings owned by low-income people. Upon paying the property tax, any city resident can choose to purchase insurance cover. Once 30 per cent of insurable buildings in the city are included, insurance cover extends to the properties of low-income people. The municipality is responsible for collecting premiums, keeping a small handling fee, whilst the insurance company takes responsibility for claims and has a direct contractual relationship with the insured individuals (Hardoy et al. 2011; Marulanda et al. 2010). Despite increasing interest in disaster insurance and other risk financing measures (see Box 2.4), risk financing is seldom adequately considered in social housing finance mechanisms (such as governmental and nongovernmental subsidies, micro-credits and family savings, mutual help or self-help mechanisms), which would be of particular significance in marginal settlements (see Table 4.6).

Apart from formal insurance policies for risk financing, in the southern context a range of informal insurance systems can be found, such as community emergency funds together with advisory services for reconstruction. Such measures are mostly implemented by NGOs but are partially also carried out in cooperation with city authorities. Finally, some few municipal (and more commonly national) authorities have developed guidelines on how citizens can quickly and safely recover. Examples are a brochure issued by the Brazilian Ministry of Health that describes a number of safety routines related to flood preparedness and recovery (SVS 2011; see Box 5.3) and similar guidelines offered by the Federal Emergency Management Agency (FEMA) of the United States via its web portal.[10]

Finally, especially in the European context, increasing importance is given to the identification of risk reduction and adaptation measures that may also reduce greenhouse gas emissions or, conversely, the identification of measures for climate change mitigation that, at the same time, may also assist in reduction of and adaptation to climate change impacts (see Table 4.6). The recent buzzword is 'climate planning' (Davoudi et al. 2009), which is the combined domain of climate change mitigation and adaptation. This development must be understood in the context of planning practices that have in many high-income nations at first focused on climate change mitigation (a different situation to that found in low- and middle-income nations). The 'adaptation turn' of planners is here, in fact, very recent (Davoudi *et al.* 2009). Initiatives for climate proofing are considered to be so-called win–win or no-regret measures (see Box 3.6). A typical win–win or no-regret measure is reforestation (or afforestation), which can reduce the risk of landslides, floods, droughts and windstorms, increases sequestration of CO_2 emissions and helps to preserve biodiversity. In addition, growing biomass can be considered a renewable future alternative to fossil fuels (European Commission 2009). Other synergies between adaptation and climate change mitigation can be found in areas such as the construction of buildings and infrastructure, which can be designed to be both disaster-resistant *and* energy efficient, in addition to keeping heat-enhancing emissions low.

In sum, analysis of the currently predominant measures for urban risk reduction and adaptation indicates that there is a huge variety of existing activities from which others can learn. Tables 4.1–4.4 provide an overview of existing measures.[11] Nevertheless, only a few cities address urban risk holistically by combining a wide range of possible measures. The identification *and* combination of adequate location-specific measures for risk reduction and adaptation, which are based on adequate risk assessments, seems to be the key challenge (see Sections 4.2–4.3 and Table 4.5).[12] More integral approaches can be found in single programmes but seldom at the city scale. City authorities thus do not, as yet, tap into their full potential. This would require a more comprehensive approach for risk reduction and adaptation that (a) combines measures which address all types of contributing risk factors, (b) considers both physical and non-physical aspects of the urban fabric and their interlinkages, and (c) builds on citizens' adaptive capacities and needs (see Chapters 2 and 3).

4.2 Overview of prevailing mainstreaming strategies

In order to gain an overview of current practice in urban risk reduction and adaptation, it is not sufficient to analyse only the predominant measures that are taken at the local level (i.e. the household, community or programme level presented in Section 4.1). It is also necessary to assess city authorities' level of risk reduction and adaptation mainstreaming, i.e. the strategies they take to sustainably incorporate risk reduction and adaptation into urban planning practice at different levels. If resilient cities are the aim, risk reduction and adaptation need to become an inherent part of urban planning, which requires the use of a set of mainstreaming strategies (see Figure 4.1, Boxes 2.8–2.10 and Table 2.7). Before continuing to read, ask yourself: what are the most common mainstreaming strategies pursued? And do you think that there are differences in city authorities' level of mainstreaming in low-, middle- and high-income nations?

In spite of the large variations that exist between cities and countries, many patterns can be identified. They are presented in the following text and are summarized in Table 4.9.[13] Note that the differences between city authorities' mainstreaming strategies in low-, middle- and high-income nations are further discussed in Section 4.3.

The first observation is that there is today a widespread consensus that risk reduction and adaptation should be presented as a mainstreaming or cross-cutting issue for urban planning, and not as an additional or separate aspect which should overrule other planning issues (see Table 4.9). This observation is independent of the specific (southern or northern) context and is a major breakthrough (see Box 2.5). Whilst risk reduction has a longer history of being recognized as a mainstreaming issue, climate change adaptation has been closely following the same development. In climate change-related literature it is increasingly emphasized that embedding essential adaptation actions into existing and future programmes, plans and policies helps to prevent adaptation being regarded as an 'optional extra' (e.g. Ribeiro *et al.* 2009: 14).

With risk reduction and adaptation being recognized as cross-cutting issues, generic aspects for stimulating mainstreaming are increasingly discussed, with great emphasis being given to top-down processes at policy level. This is especially true for high-income countries. With the emphasis on top-down policy-level interventions, particular adaptation measures or specific planning tools are generally of less interest in current debates, which is a major frustration for planners who struggle to 'translate' the 'high-level talk' to their day-to-day work (Carmin *et al.* 2012; Greiving and Fleischhauer 2012; UNISDR 2012c).

The most frequently pursued activities thus seem to be the modification of policies and regulations at different levels, which is part of organizational mainstreaming (see Box 2.9 Strategy III). Focus is here on establishing adaptation strategies and the revision of regulations for urban development, land-use planning and the construction of buildings and infrastructure (see Tables 4.7 and 4.9). During recent years, more and more countries have established national adaptation strategies, especially high-income countries. Finland was one of the first countries to adopt a National Climate Strategy in 2001 and a National Strategy for Adaptation in 2005 (see European Commission 2009). Whilst national strategies are very different in terms of their format, goals and the stakeholders involved, most do not provide sufficient guidance on aspects that are crucial to achieve mainstreaming. Such aspects include the definition of different actors' responsibilities; working structures and processes to support continuity for the implementation processes; cost estimates and availability of required resources; the incorporation of risk reduction and adaptation into governmental and municipal budgets; and the revision of operational tools – ranging from risk assessment to planning and systematic monitoring and reporting.[14] As regards this last, in the face of complex and uncertain predictions of climate change impacts, policy integration at all levels must be able to learn from past experience, requiring careful pre-assessment and post-evaluation of risk reduction and adaption measures and the monitoring of the effectiveness of related policies (Mickwitz *et al.* 2009; Ribeiro *et al.* 2009; Swart *et al.* 2010). In this effect, the National Indicator 188 has been developed in the UK, which reports on the preparedness of local authorities to deal with climate change adaptation. It is one of 198 indicators of the Local Government Performance Framework introduced in 2008, which is currently under revision (CoR 2011).

Examples of the revision of existing regulations (such as building codes and urban development regulations) are as numerous as the development of new regulations. A case in point is the 'Planning Guidance on Flood Risk and Development' of Salford City, UK. It was adopted in July 2008 on the basis of the flood level predicted for a 1 in 1,000-year extreme flood event.[15] An example at European level is the EU road safety directive (Directive 2008/96/EC), which was revised to ensure that infrastructure standards consider possible exposure to hazards, including seasonal and climatic conditions (Jürgens 2011). For infrastructure planning, mainstreaming is considered particularly crucial due to the urban fabric's long lifetime of between 40 and 100 years. It means that if climate change is not considered in today's infrastructure planning, the opportunity of a century is lost. Regardless of this fact, even in cases where infrastructure planning has been adjusted to withstand a 100-year event, related modifications are generally only based on current climate conditions and exclude future climate projections (Jürgens 2011).

Whilst the revision of policies and regulations generally aims at better integrating risk reduction and adaptation, there are also examples that are meant to hinder related improvements as shown in the example of North Carolina (Box 4.3).

BOX 4.3

North Carolina – trying to outlaw sea-level rise

In 2012, a group of legislators from 20 coastal counties in North Carolina in the United States adopted a new institutional strategy against sea-level rise by making its prediction against the law (*The Guardian* 2012). According to the State General Assembly's Replacement House Bill 819, future scenarios for sea-level rise shall only be based on historical data and extrapolated in a linear fashion, i.e. ignoring the evidence and global consensus that sea levels are rising at an accelerated speed due to global warming. The 'permitted', linear calculations predict that sea levels will only have risen about 20 cm by the year of 2100, instead of 1 metre (which is the forecast based on an exponential increase). The State General Assembly's Replacement House Bill 819, Section 2, Paragraph E reads:

> The Division of Coastal Management shall be the only State agency authorized to develop rates of sea-level rise and shall do so only at the request of the Commission. These rates shall only be determined using historical data, and these data shall be limited to the time period following the year 1900. Rates of sea level rise may be extrapolated linearly to estimate future rates of rise but shall not include scenarios of accelerated rates of sea-level rise.[16]

In contrast to high-income nations, cities in middle- and low-income nations generally face more difficulties in integrating risk reduction and climate change concerns into policies and regulations, mainly due to a lack of resources and institutional capacities (UNFCCC 2007). But there are exceptions. In Cape Town, South Africa, the city authority has mandated that the Municipal Disaster Risk Management Centre has to be involved in the review process of all new development projects (see Box 4.4).[17] In Peru, a pioneering legal requirement has been established, which only allows the funding of public investment projects if disaster risk is addressed (UNISDR 2011b). Changes in policies and regulations are, however, of little impact in cities that are characterized by a lack of enforcement control and informal settlements that develop outside any official plans and regulations. In many southern cities, legal documents for land-use planning can further become a question of moral and human rights because it is not justifiable to put up barriers to prevent people from settling in risk areas if such interventions expose them to even greater economic insecurity. Such circumstances require city authorities to deal with informal areas *in situ* to encourage practices and structures that reduce risk without depriving people of their assets. This is, however, rarely done in practice.

In contrast to the revision of policies and regulations at different levels, less consideration is generally given to the actual modification of the organizational management, working structures, budgeting and tools of urban actors, which is also part of organizational mainstreaming (see Tables 4.7 and 4.9). The most common changes in organizational management and working structures are probably the implementation of interdepartmental focal points, task forces and commissions, specialized departments or other advisory groups, as well as the increase in the number of staff for risk reduction and adaptation planning (see Table 4.7). A global survey on urban climate adaptation planning found that 54 per cent of the 468 responding cities had formed some type of group to support adaptation planning; notably, the corresponding figure for Latin America was over 75 per cent (Carmin *et al.* 2012). An example of a specialized department is 'Makati Command, Control and Communication' (Makati C3) of Makati City, the Philippines, which was established in 2006. In accordance with its mandate to increase disaster resilience, it takes an active role in supporting risk-sensitive land-use planning, community-based risk reduction and related capacity building programmes (UNISDR 2012b). As regards the number of staff specializing in risk reduction and adaptation planning, at Copenhagen municipality in Denmark, for instance, it was increased from two to eight in 2012 (Leonardsen 2012). The temporary employment of new staff is common when their payment comes from (often) external and time-limited sources – as opposed to the increase in permanent staff, their capacity development and the revision of their job descriptions.[18] Nevertheless, there is surely also increasing interest in initiatives (by international and national actors) to develop capacities of municipal staff and planners, although knowledge on capacity development for risk reduction and adaptation is still generally low (Hagelsteen and Becker 2013).

As regards the financing of risk reduction and adaptation measures and its incorporation into existing budgeting, cities face several challenges. Municipal budgets are generally the primary source of financing (UNISDR 2012c). Other sources are not easy to access because of (a) a lack of knowledge on existing funding sources, (b) personal and administrative difficulties in articulating and initiating requests and (c) institutional mandates or unclear legal responsibilities resulting in so-called 'catch-22 situations' (i.e. no-win situations). As regards the first point, many cities are struggling with how to access international climate change funding, especially in low- and middle-income nations (UNHABITAT 2011). Meanwhile, the latter can be illustrated with the case of Copenhagen City. Stormwater management is here financed by water companies, whilst the municipality is responsible for the road network and for improving citizens' quality of life. If these aspects are combined in multifunctional measures (e.g. green solutions for water management or special street design for improved water management), their financing becomes hardly possible (Leonardsen 2012). Another example is money that has been earmarked for water management and which cannot be used to divert water in underground reservoirs to lead the water out to the sea, although this would avoid the expansion (and related financing) of the city's sewage system (Leonardsen 2012).

In relation to the modification of operational tools to account for risk reduction and adaptation, most advances can be observed in the context of risk assessment, which is a necessary precondition for all types of risk reduction and adaptation measures (see Box 2.4, Figure 2.6 and Table 4.7).[19] Related advances can particularly be observed in the northern hemisphere. This is, amongst other issues, related to an increase in mandatory regulations to elaborate, improve and harmonize national and municipal risk assessments. An example is Directive 2007/60/EC, which demands EU member states to map and assess flood risk areas and resources.[20] In Sweden, as of January 2011, municipalities and city councils are required by the Swedish Civil Contingency Agency (MSB) to evaluate and abridge their work on extraordinary events, present their outcomes every year and provide a full risk assessment every fourth year (MSB 2012a). As a result, the use and harmonization of risk assessments is increasing. Risk assessment is also often the focus, or an important part, of new tools elaborated to support city authorities and aid organizations in their efforts to mainstream risk reduction and adaptation. Examples are the *Tools for mainstreaming disaster risk reduction* (by Benson *et al.* 2007), the *Toolbox for climate change adaptation* (by Jonsson *et al.* 2011), *CLIMATOOLS* [21] and the *Local Climate Impacts Profile* (LCLIP).[22] Furthermore, there is an increasing number of more general guides for improved adaptation planning, especially in the context of urban water management (e.g. Jha *et al.* 2012; Loftus *et al.* 2011; Shah and Ranghieri 2012). Despite the improvements in tools and guidelines for risk reduction and adaptation planning, many challenges persist: current measures for risk reduction and adaptation are seldom based on actual risk assessments (see Carmin *et al.* 2012; Table 4.6); implemented risk assessments generally do not give adequate (or any) consideration to the analysis of the city–disasters nexus and citizens' adaptive capacities (described in Chapters 2 and 3); and the general increase in tools seldom translates into actual practice (which is related to the identified gap between research and policymaking; see Table 4.10). In the southern hemisphere, further challenges persist at city and national levels (see Section 4.3).

Apart from the above-described progress in organizational mainstreaming, many advances relate to the enhancement of interinstitutional cooperation, synergy creation and the harmonization of risk reduction and adaptation to improve current risk governance structures (which is part of mainstreaming Strategy V; see Box 2.10 and Table 4.8). The importance given to these issues is underpinned by international policy documents and international campaigns (such as the Hyogo Framework for Action, which promotes the establishment of national platforms for risk reduction [see Box 2.7] or the 'Making Cities Resilient' Campaign which enhances city-to-city learning [see Box 2.12]). It is further based on a general consensus that the successful implementation of risk reduction and adaptation planning is highly dependent on the level of commitment and leadership of local authorities, and their interaction with other urban stakeholders. In other words, weak risk governance capacity is commonly understood to be a root cause of the current failure in risk reduction and adaptation planning. To improve existing risk governance systems, high-income nations put most emphasis on enhancing the coordination and cooperation between different urban authorities (see Boxes 2.12–2.13) and creating public–private partnerships. To improve coordination and knowledge sharing, national or regional platforms and several online portals have been established for city authorities, such as *UKCIP* in the UK,[23] *Stadtklimalotse* in Germany,[24] *Klimatanpassningsportalen* in Sweden[25] and *Klimatilpasning* in Denmark.[26] Citizens are, however, rarely considered in the strategies to advance current risk governance structures. In low- and middle-income nations, a comparatively greater involvement of NGOs can be observed in pushing forward the improvement of current risk governance structures, which also leads to greater importance being given to civil society organizations and citizens at risk (see Table 4.9). Although the stronger involvement of NGOs, community-based organizations (CBOs) and citizens is mainly due to the lack of functioning governmental structures, their experience in risk reduction and adaptation planning is manifold and provides a rich information source. This includes lessons learnt on how to actively involve and work with citizens at risk, from planning to implementation, monitoring and evaluation, and on how to upscale community-based responses to municipal levels (for instance by linking them to city-wide responses) (see Section 4.3).

Unlike organizational and interorganizational mainstreaming, programmatic, internal and educational mainstreaming are generally given little attention by city authorities (see Boxes 2.8–2.10 and Tables 4.7–4.9). Programmatic mainstreaming requires local knowledge on citizens' differential vulnerability and the complex links between the urban fabric, climate change and disasters, which is not available to most city authorities. Adding to their core work some 'simple' activities that are explicitly aimed at reducing risk is often the easiest way out when actions in risk reduction and adaptation are demanded (see Table 4.9). This approach fails, however, to consider the basic mainstreaming strategy of responding indirectly through planners' normal day-to-day operations.

Furthermore, there are generally few modifications to city authorities' internal functioning to ensure the protection of staff and the organization in case of disaster. This issue is seldom discussed, despite the many disaster impacts that affect city authorities themselves, leading to costs for clean-up, repair, lost employee labour and disruption of services (e.g. due to lost files or damage to communications and computer equipment). Damage to non-structural elements can be a major cause of organizations' loss of functionality. A range of potential measures for internal mainstreaming can be found in guidebooks for private enterprises or health facilities which aim to assist organizations in offering their services sustainably – often under the buzzword 'benchmarking resilience' (e.g. Stephenson et al. 2010; World Bank 2005). A Hospital Safety Index was, for instance, established by the Pan American Health Organization to help health facilities assess their level of risk and avoid becoming a casualty of disasters.[27] By determining a Hospital Safety Index, decision-makers gain an overall idea of the hospital's ability to cope with disasters and it is an important first step towards prioritizing investments in hospital safety. Similar assessments are seldom carried out for city authorities.

Potential internal mainstreaming measures for city authorities are listed in Table 4.7. They include issues such as the protection of the organization's infrastructure and equipment, staff security, access to disaster information, changes to make existing budget lines disaster/climate-proof and establishing back-up communication plans in case any phone, road or postal communications are cut off. This is crucial to ensure an organization's continuous functioning, also in times of increasing disasters and climate change. The 2010 earthquake in Chile is a good example to illustrate this. The lack of adequate and transparent back-up plans for internal communication was one of the major shortcomings that impeded a timely post-disaster reaction and cost many people's lives. Existing information could not be effectively shared between national and municipal organizations. Of 14 satellite telephones only 4 had a chip and 2 were missing. The issue has reached the highest political level and has recently been brought to court (Ayala 2012; BBC Mundo 2012).

As for educational mainstreaming, although there seems to be greater momentum towards closer cooperation between city authorities, universities or other educational bodies for mutual learning, in practice it is still scarce (see Tables 4.8–4.9). In fact, educational and research institutions only have a limited role in actual practice, and city authorities and professional planners have little influence on research agendas, curricula development and education (UNISDR 2011a). Existing cooperation between city authorities, universities and other educational bodies can involve minimal engagement (such as the unidirectional provision of information, the interpretation of academic or scientific information into recommendations for policy and management options, or students' assignments to support municipal work on urban risk reduction and adaptation).[28] Cooperation can (and should) also involve more active engagement (including a two-way knowledge transfer). Positively, research projects increasingly include components committed to communicate research outcomes to decision-makers. This is, for instance, the case of the research project CIRCLE-2 (Climate Impact Research and Response Coordination for a Larger Europe).[29] Illustrative examples of mutual collaboration between city authorities, educational and research institutions are the Lund University Centre for Sustainability Studies (LUCSUS), which is collaborating with Lund municipality and other city authorities,[30] the Institute for Sustainable Urban Development (ISU), which is a joint venture between the city of Malmö and Malmö University (EEA 2012),[31] the University of Manchester, which has a close working relationship with the city council (CoR 2011), and the Institute of Environmental Studies (IDEA) of the National University of Colombia in Manizales, which works closely with the city of Manizales on integrating risk reduction and adaptation into urban planning practice.[32]

In sum, the analysis of the current mainstreaming strategies taken by city authorities and other urban actors shows a wide range of possible actions to make risk reduction and adaptation a standard procedure of urban planning practice (see Tables 4.7–4.8); however, mainstreaming is seldom carried out in a comprehensive way and is left to single actions. Most emphasis is given to the revision of existing policies and regulatory frameworks, followed by the improvement of interinstitutional cooperation and communication structures to improve risk governance. Academic bodies and citizens are, however, rarely considered in efforts to advance current risk governance. Furthermore, little consideration is generally given to the revision of city authorities' organizational management, working structures, financial

Table 4.7 Mainstreaming strategies taken by city authorities with focus on their organizational functioning

Strategies	Illustrative examples of activities
Organizational mainstreaming (Strategy III)	Revision of existing, or establishment of new, regulations and policies for urban development, land-use planning and the construction of buildings and infrastructure so as to account for increasing disasters and other climate change impacts
	Establishment of new legal requirements to assess risk and evaluate urban development work as regards its effects on risk reduction and adaptation (e.g. establishment of legal requirements only allowing the funding of public investment projects if disaster risk is addressed)
	Inclusion of the commitment and responsibility of the organization and its different departments to respond to increasing disaster risk and climate change in key (guidance) documents of the organization (e.g. mission statements, strategy papers, establishment of a municipal adaptation strategy)
	Establishment of interdepartmental focal points (task forces, commissions, specialized departments or other advisory groups) for the integration of risk reduction and adaptation into all kinds of urban sector work
	Establishment of specialized departments for construction control (to assure the disaster resistance of buildings and infrastructure)
	Increase in the number of staff for (mainstreaming) risk reduction and adaptation planning
	Capacity development of staff in (the mainstreaming of) risk reduction and adaptation
	Modification of job descriptions to include employees' responsibility for considering disaster risk and its reduction within their daily practice
	Revision of existing planning tools and development of additional tools (e.g. risk/hazard assessment) to account for increasing disasters and other climate change impacts
	Definition of different actors' responsibilities, working structures and processes, estimates and availability of required resources (both in terms of costs and know-how), mechanisms to support continuity of the implementation processes and systematic reporting
	Design and implementation of a financial strategy for (mainstreaming) risk reduction and adaptation
	The establishment and/or use of indices that can assist in prioritizing a city authority's investment for and engagement in risk reduction and adaptation (e.g. Hospital Safety Index)

(continued)

Table 4.7 (continued)

Strategies	Illustrative examples of activities
Internal mainstreaming (Strategy IV)	Disaster-resistant construction or upgrading of the organization's office buildings
	Construction of fireproof or waterproof cabinets within the organization's office buildings
	Storing (copies of) key documents in a secure offsite location
	Considering disaster insurance for office facilities and equipment
	Making existing budget lines disaster/climate proof
	Providing health, life and/or damage insurance for staff
	Development of policies on assistance to disaster-affected staff members
	Development of a policy on 'disaster-watch' and establishment of channels to disseminate warnings (e.g. establishment of an early warning communication chain among employees)
	Ensuring a back-up communication plan in case any phone, road or postal communications are cut off
	Identifying institutions (e.g. government, hospitals) with access to radio or satellite communication, which could assist during a natural disaster
	Development of a policy on vehicle and other equipment evacuation, and training of staff on these policies
	Assuring a back-up power supply through the pre-purchase of a generator and/or batteries
	Development of a staff evacuation plan and related staff training (simulations)
	Identifying alternative access routes to clients, and if necessary establishment of means for alternative transportation

Table 4.8 Mainstreaming strategies taken by city authorities with focus on capacity development and improved science–policy interface

Strategies	Illustrative examples of activities
Interorganizational mainstreaming (Strategy V)	Establishment of national platforms for disaster reduction and adaptation
	Establishment of online portals to provide support for adequate coordination and knowledge sharing for risk reduction and adaptation planning
	Establishment of better communication structures and mechanisms between different actors responsible for risk reduction and adaptation
	Creation of public–private partnerships
	City-to-city learning and exchange programmes
	Activities to address existing competition between different organizations, political manoeuvring or a lack of institutional cohesion (e.g. establishment of definitions of roles and responsibilities of different actors)
	Support of participatory risk reduction and adaptation involving at-risk citizens (e.g. use of micro planning or strategic action planning, scenario workshops and citizen summits, see Section 4.3)
Educational mainstreaming (Strategy VI)	Cooperation with universities to develop operational tools that can assist municipalities in their work on urban risk reduction and adaptation
	Cooperation with universities for the interpretation of academic or scientific information and its use for practice (e.g. their 'translation' into policy and management options)
	Involvement of municipal staff and professional planners in research agendas, curricula development and (undergraduate and postgraduate) teaching
	Support of students' assignments or other research on city authorities' work on urban risk reduction and adaptation
	Two-way knowledge transfer between city authorities and educational/ research institutions (e.g. joint implementation of projects)

mechanisms and operational planning tools to ensure the sustainable integration of risk reduction and adaptation at programme level. Exceptions to this are the establishment of interdepartmental focal points, task forces and commissions, specialized departments or other advisory groups as well as the promotion and harmonization of risk assessment tools. Finally, little consideration is given to the risk faced by city authorities themselves, even when headquarters or staff are located in high-risk areas. There are thus hardly any city authorities that show a high level of risk reduction and adaptation mainstreaming (see Chapter 2, Boxes 2.8–2.10).[33] The engagement in all mainstreaming strategies is, however, crucial even for those city

Table 4.9 Similarities and differences between city authorities' mainstreaming strategies in low-, middle- and high-income nations

Context	Mainstreaming strategies – implementation and integration processes
Similarities: not context specific	Risk reduction and adaptation are seen as mainstreaming issues for planning, not as separate issues The predominant mainstreaming strategy is organizational mainstreaming: focus here is on modifying policies and regulation. Little attention is given to the revision of organizational management, working structures, financial mechanisms and operational planning tools. Exceptions are the establishment of interdepartmental focal points, task forces and commissions, specialized departments or other advisory groups and the promotion and harmonization of risk assessment tools Second in importance is interorganizational mainstreaming to improve risk governance structures and the communication between the different stakeholders (e.g. by engaging in multi-stakeholder consultations) The implementation of separate add-on risk reduction and adaptation measures is often given more consideration than the modification of planners' day-to day work (i.e. programmatic mainstreaming) Programmatic, internal and educational mainstreaming is given little consideration (in comparison with organizational and interorganizational mainstreaming)
Differences: specific to high-income countries	More top-down approach, but increasing interest in participatory methods (see Section 4.3) Strong focus on the implementation process of adaptation mainstreaming, whilst particular measures or specific planning instruments are of little interest Interinstitutional mainstreaming: focus on cooperation and communication between urban authorities to improve risk governance; little involvement of NGOs, civil society organizations and citizens Organizational mainstreaming: the majority of related activities is focused on the revision of land-use planning and urban development regulations
Differences: specific to low- and middle-income countries	Interinstitutional mainstreaming: focus on cooperation between urban stakeholders, with greater involvement of NGOs, civil society organizations and citizens to improve risk governance structures Organizational mainstreaming: related action-taking might have little impact in cities that are characterized by informal settlements and a lack of resources and institutional control capacities (also resulting in non-compliance of regulation in formal areas) Result: more community-based and bottom-up approaches and different techniques and methods taken at the household level, and less involvement of planners and formal planning processes, especially in informal areas (see Section 4.3)

authorities that are today well prepared for 'conventional' weather extremes (such as riverine flooding) because the systems and mechanisms associated with these extremes are often 'outdated' in the context of climate change.

An example of a city authority that is engaged in all strategies necessary to achieve a sustainable mainstreaming of risk reduction and adaptation is Manizales in Colombia, which is well known for its integrated approach (Hardoy *et al.* 2011).[34] Manizales is a good case to illustrate how the different mainstreaming strategies and measures for risk reduction and adaptation can be combined to support each other. The following is a description of the Manizales case, including an analysis of the employed strategies and measures (see Boxes 2.4, 2.8–2.10):[35] the city authority's work on risk reduction and adaptation integrates local and regional governments, the private sector, universities and representatives of community organizations into a participative process (*Strategy V and VI: Interorganizational and Educational Mainstreaming*). Legislation supports and strengthens this approach (*Strategy III: Organizational mainstreaming*). As an example, the Urban Planning Law (*Ley de Ordenamiento Territorial 1999*) requires all urban development plans to be validated by local planning committees and involve the participation of civil society organizations, educational bodies, research institutions and a range of other urban actors. In addition, planning mechanisms and regulatory frameworks were set up to support the mainstreaming of risk reduction and adaptation into urban sector work. The Municipal Disaster Prevention Strategy requires that risk reduction is an integral part of local policies, and the city's urban development plan is required to integrate the municipal disaster risk management plan. This has also led to the adoption of new tools for urban development planning, such as risk mapping and micro-zoning (*Strategy III: Organizational mainstreaming*). Furthermore, environmental observatories and a simple set of indicators on the environmental conditions in the city's 11 districts were established to monitor progress made, including progress in the reduction of disaster risk (*Strategy III: Organizational mainstreaming*). These indicators are represented visually by environmental 'traffic lights' (*semáforos ambientales*) and have been operating as a communication and public awareness tool (Figure 4.4). The work with the indicators is a continuous process, which strongly involves the citizens of Manizales (*Strategy V: Interorganizational mainstreaming*) and led to widespread support in implementing the city's environmental plan (*Strategy III: Organizational mainstreaming*). The integration of risk reduction into urban planning has also benefited from the direct support of academic and research institutions, such as the Institute of Environmental Studies of the National University of Colombia in Manizales (*Strategy VI:*

Educational mainstreaming). Furthermore, some of the municipal buildings have been retrofitted to become more disaster-resistant, internal emergency response plans are distributed to municipal staff and related simulations are carried out on a regular basis (*Strategy IV: Internal mainstreaming*). The analysis of the measures taken at local household level shows that risk reduction has become part of planners' day-to day work in many ways (*Strategy II: Programmatic mainstreaming*). The identification and design of risk reduction measures is generally based on risk assessments and the direct involvement of citizens (*Strategy V: Interorganizational mainstreaming*). Concrete examples are the work with citizens to relocate them to safer sites and the conversion of the 'cleared' risk areas into neighbourhood parks (*Hazard avoidance, combined grey and green measures, direct risk reduction*).[36] In addition, slopes are stabilized (*Hazard reduction, use of grey and green measures, direct risk reduction*) and the city authorities have established the slope guardians programme involving more than 100 female heads of households who live where they work (*combined hard and soft measures*). They receive training and, in turn, work on raising awareness, maintaining and controlling slope stabilization works, reporting problems and communicating experiences to other communities (*Hazard reduction, soft measures, capacity development*). For the slope guardians programme, a support team was created consisting of professionals and technicians from the municipality of Manizales, the regional environmental entity (CORPOCALDAS), the Red Cross, the local water company (*Aguas de Manizales*) and the Institute for Environmental Studies (IDEA) from the National University (*Strategies V and VI: Interorganizational and educational mainstreaming, soft measure*). Furthermore, the city authority has revised the tax system (*Strategy III: Organizational mainstreaming*) to include several innovative financing mechanisms, such as tax reductions for those who take measures to reduce the vulnerability of their housing at risk of landslides and flooding (*physical vulnerability reduction, soft/economic measure, incentives for risk reduction*) and an environmental tax on rural and urban properties that is reinvested in environmental protection infrastructure, community education and relocation of at-risk communities (*Hazard reduction and avoidance, combined green, grey and soft/economic measures, direct risk reduction and capacity development*). The city has also established an insurance programme for buildings owned by low-income households (*Strategy III: Organizational mainstreaming [new tools]; preparedness for recovery*). Once 30 per cent of insurable buildings in the city participate, insurance coverage extends to low-income households. The municipality collects premiums, keeping a small handling fee, but the insurance company takes responsibility for claims and has a direct contractual relationship with the insured individuals (*Strategy V: Interorganizational mainstreaming*). Finally, the municipality is also well prepared to respond to potential disasters (*Preparedness for response*): simulations are carried out on a regular basis, the municipality uses predefined procedures for evaluating the safety of damaged buildings and the demolition of houses in danger of collapse (*Strategy III: Organizational mainstreaming*), and inter-institutional agreements are established between the municipality and different relief agencies to ensure adequate disaster response (*Strategy V: Interorganizational mainstreaming*).

4.3 Differences in institutional approaches: current planning practice and increasing risk

Historically, one of the main functions of cities was to provide protection and defence for their inhabitants. City planning has thus always been an act of risk reduction and adaptation. However, due to the emergence of more and more threats, and the inadequate management (or neglect) of cities' protective and defensive function, many cities are today characterized by increasing risk and disasters.

Whilst national, regional and municipal planning systems can vary greatly in terms of their priorities, the scope and extent of their powers, their regulatory frameworks and the resources with which they work, there are also striking similarities in current planning practice, even when comparing low-, middle- and high-income countries. Before continuing to read, ask yourself: what are the key challenges city authorities face when planning for risk reduction and adaptation?

Many city authorities and planners around the world are faced with the challenges listed in Table 4.10, all of which can hamper risk reduction and adaptation and its integration into urban planning practice (see Section 2.4). These challenges relate to (a) city authorities' organizational setting, (b) their technical planning bases, (c) their cooperation with, and support from, other urban actors and (d) other contextual aspects (e.g. CoR 2011; Carmin *et al.* 2012; UNISDR 2012c).

City authorities' organizational setting can hamper (the integration of) risk reduction and adaptation which relates to aspects such as their human and financial resources, political support, organizational mandates, working structures and control mechanisms (see Table 4.10). City authorities often cite the lack of adequate human and financial resources as the major barrier to their work on risk reduction and adaptation. Since organizations are made of people, the mainstreaming of risk reduction and adaptation into an organization is dependent on their continuous commitment and motivation, which requires staff 'ownership' of this issue. There are, however, several barriers to staff 'ownership', including fear of increased workload, lack of institutional leadership, non-participatory decision-taking and lack of skills and knowledge to understand the urgency of risk reduction and adaptation and its relevance to people's own work (La Trobe and Davis 2005). Another barrier is the general lack of human and financial resources, together with little political support to control the construction sector (which can result in the use of substandard and unsafe building materials or techniques and the colonization of risk areas).

Another issue which is related to the control capacity of the construction sector is the politicization of urban planning (see Table 4.10). It can translate into non-compliance with regulations, such as risk assessments prior to new urban developments. The resettlement of disaster-affected people in (other) high-risk zones is a common consequence. A recent example comes from Rio de Janeiro where families who moved out of risk areas were offered social housing in *Bairro Carioca*, developed in collaboration between the municipality and the federal government. Shortly thereafter, the new neighbourhood was flooded. Most likely, this situation occurred due to the speed with which the area was developed in order to hand out the apartments before the election period (O Globo 2013).

As for city authorities' technical planning bases, they can hamper risk reduction and adaptation planning due to lack of or outdated data and maps, as well as imported regulations that do not consider local hazards (see Table 4.10). An example is the common use of outdated, imported or colonial building and planning regulations that do not consider context-specific factors such as local hazards, as well as the use of 'best local practices' for the design and construction of infrastructure which do not consider relevant considerations for hazard-resistance, let alone climate change.

The lack of cooperation with or support from other urban actors at different levels (including citizens) presents another challenge (see Table 4.10). It results in weak risk governance capacity, which is commonly understood to be a root cause of the current lack of comprehensive risk reduction and adaptation planning (see Section 4.2).

Finally, it is the rapidly changing context and related uncertainty that can hamper urban risk reduction and adaptation caused by aspects such as increasing disasters, climate change and urbanization (see CoR 2011). The rapid modification of the urban fabric makes many city authorities unable to offer residents adequate services and protection (Table 4.10).

The above-described challenges have also been identified in recent research on Swedish city authorities' perceived barriers to adaptation (Jonsson *et al.* 2010). The interviewees' answers to the question 'What affects your adaptive capacity?' were summarized under the following eight factors: lack of resources, lack of political support, goal conflicts, unclear responsibilities, deficient support material, lack of coordination, lack of public awareness and inability to track global changes – all of which form part of the aspects listed in Table 4.10.

Finally, an additional barrier to mainstreaming is the concept itself. The mainstreaming concept is widely misunderstood and there is a lack of related knowledge and tools that are relevant and applicable at city and local levels (see Carmin *et al.* 2012; UNISDR 2010a; Wamsler 2009a; Pelling 2007; SKL 2011). One of the aims of this book is to counteract this situation by providing a conceptual and operational framework for mainstreaming risk reduction and adaptation into urban planning practice (see Section 2.4). This framework was designed to address the different challenges and barriers city authorities face when planning for risk reduction and adaptation (see Table 4.10).

Before continuing to read, ask yourself: how are the challenges presented in Table 4.10 reflected in the current measures and strategies taken by city authorities (presented in Sections 4.1–4.2)?

There are many similarities in the predominant measures and strategies taken by city authorities around the world. Many trends and developments are similar and are listed in Tables 4.6 and 4.9. But there are also strong differences. They relate to location-specific variations in the challenges listed in Table 4.10, cities' level of exposure and the existing institutional capacities, which lead to a stronger or lower involvement of NGOs, CBOs and citizens. This translates, in turn, into quite different scenarios for risk, risk reduction and adaptation on the ground, which become especially striking when formal settlements in high-income nations are compared with informal (slum) areas in low-income nations (see Tables 4.6 and 4.9).

Urban risk reduction and adaptation in low- and middle-income nations can often be characterized by an increased use of:

- Temporary solutions (e.g. plastic sheets to cover slopes and divert water)
- Low-tech measures (for instance for early warning, see Box 4.2)
- Small-scale physical measures (e.g. appliances to lift objects during floods, buried waterproof containers designed to hold drinking water

Table 4.10 Challenges faced by city authorities and planners that hamper risk reduction and adaptation and its integration/mainstreaming into urban planning practice

Key challenges and constraints[a]	Examples of specific challenges and constraints
City authorities' organizational setting: • human and financial resources • political support and politics • organizational mandates • working structures • control mechanisms	Lack of financial resources in general. For instance, decentralization of planning functions without decentralized technical and financial resources (e.g. due to structural adjustment processes). This results, for instance, in the erosion of living standards (e.g. through poor maintenance of rental property and funds being unavailable for housing and related risk reduction)
	Lack of staff 'ownership' of risk reduction, adaptation and their mainstreaming due to, for instance, fear of increased workload, lack of institutional leadership, non-participatory decision-taking and lack of skills and knowledge to understand the urgency of risk reduction and adaptation and its relevance to people's own work
	Insufficient human resources in general and in particular with expertise in risk reduction and adaptation
	Lack of wider institutional backing for sustained impact of so-called risk reduction or climate change 'champions' within urban institutions (although being potentially key catalysts for driving agendas)
	Lack of specific funding/budget lines for risk reduction and adaptation
	City authorities' lack of political support and financing, for instance, to control the construction sector (e.g. corruption leading to the use of substandard building materials and inability to force developers and property holders to plan in secure areas and/or invest in security features)
	Politicization of building and planning processes as illustrated, for instance, by global bidding process between cities to attract investment, resulting in planning authorities being forced to lower environmental regulations (and security of workers)
	Goal conflicts: mandates to address certain issues exclusively related to their office, which reduces the amount of time and resources that can be dedicated to cross-sectoral issues, policies and programmes (see Section 4.2)
	Little interdepartmental cooperation (e.g. for physical and social planning) which is required for risk reduction and adaptation planning
	Few adequate mechanisms to finance/access safe land and disaster resistant housing for the poor
	Unclear borders between individual and institutional responsibilities to take adaptive measures on the ground
	Poor enforcement schemes resulting in non-compliance of building and planning codes
	Absent or poor certification and licensing of planning professionals (who would be responsible for applying, enforcing or inspecting codes) because of, for instance, disciplinary traditions and corruption

Table 4.10 (continued)

Key challenges and constraints[a]	Examples of specific challenges and constraints
Technical planning basis:	Lack of data: deficient data on future climate change and hazards, related impacts and vulnerabilities
• databases	Use of out-of-date and incompatible databases and maps
• maps	Inadequate regulations and policies, for instance, use of outdated, imported or colonial building and planning
• regulations and policies	regulations that do not consider context-specific factors such as local hazards
	Use of 'best local practices' for the design and construction of infrastructure, which ignore relevant considerations for hazard-resistance and climate change
	Frequent non-compliance with regulations prescribing certain methods or tools, such as the use of environmental impact assessments (EIA) for new urban developments, updates of risk assessments, etc.
Link with other urban actors:	Lack of communication and cooperation between national and municipal planning authorities
• horizontal	Few institutional linkages with disaster-related agencies in general
• vertical	Corruption at all levels, unnecessary bureaucracy and political rivalry within and between different sectors and ministries
	Competition for scarce financial resources for adaptation (especially at local and regional levels) and the highly complex way climate funding is structured through its various mechanisms at different levels
	Dependency on other agencies as regards information required for urban risk reduction and adaptation (e.g. on global changes)
	Gap between research and policymaking, which may result in a lack of appropriate understanding of risk reduction and adaptation issues among policymakers, hampering the creation of scientifically supported policies and programmes
	Time-consuming process of developing trust and building a team approach for cooperation between relevant groups and sectors
	Lack of public awareness as regards climate change, risk reduction and adaptation, which makes cooperation challenging
	Complexity of (more) distributed urban governance systems makes decision-making a lengthy process

(continued)

Table 4.10 (continued)

Key challenges and constraints[a]	Examples of specific challenges and constraints
Contextual aspects: • increasing disasters • climate change • rapid urbanization	Urban fabric's constantly changing risk patterns (see Chapter 3) Constantly growing cities lead to authorities not being able to supply adequate housing, basic services and assistance at the same speed as urbanization processes Cities growing together without effectively integrating their urban and disaster management agencies, resulting in confusion and the inability to coordinate any risk reduction measures (See Section 3.3 on disaster impacts on cities)
Mainstreaming concept	Mainstreaming concept is widely misunderstood Lack of knowledge and tools for mainstreaming that are relevant and applicable at city and local level

Note:
[a] To these challenges add aspects related to the characteristic city features (see Chapter 3), such as reduced local leadership structures, etc. (see Table 3.3).

Figure 4.3 Locally built retaining walls made of old car tyres in San Salvador, El Salvador, and sandbags filled with soil and seeds in Medellín, Colombia.
Sources: Christine Wamsler and www.inteligenciascolectivas.org

or valuables, or higher platforms for emergency refuge and rescue endeavours)
- Measures which are based on local and/or traditional knowledge
- Cheap and imaginative materials (e.g. waterproof recycled materials used to protect temporary dikes, or tyres for retaining walls [Figure 4.3])
- Innovative insurance systems such as emergency community funds or disaster micro-insurance
- Soft measures to complement other types of measures (e.g. local monitoring groups to prevent people settling on risky slopes and for maintaining retaining walls)
- People's involvement and training (e.g. training of informal builders, see also former aspect and related example) (Box 4.4).

In low- and middle-income nations more consideration is thus generally given to techniques and methods used for community-based and bottom-up approaches to risk reduction and adaptation. Community-based adaptation to climate change in urban areas has emerged from experiences in disaster risk reduction (as is the case for many approaches to deal with climatic extremes and variability; see Table 2.6 and Boxes 2.5–2.6). Examples of community-based and bottom-up techniques and methods are:

- Land pooling and readjustment: a tool whereby a group of separate land parcels can be assembled for unified planning for risk reduction and adaptation in order to then subdivide and service it as a single area. Some of the new building plots are sold to recover the costs and the other plots are redistributed to the landowners.

- Exchange of dwellings: some dwellings or single apartments are exchanged so that more vulnerable people (e.g. less mobile people) can be placed in the most secure sites. For instance, the elderly, the disabled and children are moved to higher ground within a flood area or closer to access roads.

- Exchanging rights schemes: a tool that aims to reduce population densities in high-risk areas by transferring people to zones of lower risk. If, for instance, a five-storey building has been constructed in a high-risk area, with the current law only allowing three levels, the building rights for the two additional levels can be transferred to a more secure part of the city.

- 'Barefoot planners': urban planners and architects who offer door-to-door advice on the upgrading of housing (e.g. more disaster-resistant building).

- Participatory approaches for local awareness raising and active involvement of citizens in risk assessment and reduction (see Box 4.4): examples of concrete measures are the use of environmental traffic lights in Manizales, Colombia (Figure 4.4) and citizens groups organized by the municipality to bring a wide range of stakeholder organizations together in decision-making.

- Sustainable livelihoods approach: a holistic and people-centred approach to understanding and addressing the diverse factors that influence poor people's livelihood and well-being, with the aim of enhancing their ability to support themselves in an economically, ecologically and socially sustainable way. Local analyses include the assets people draw upon, the strategies they develop to make a living, the context within which a livelihood is developed, and the factors that make a livelihood more or less vulnerable to shocks and stresses, which includes climatic and non-climatic hazards (Krantz 2001; IRP 2010a).[37]

- Micro and strategic action planning: a progression of community action planning to ensure participatory and integrated settlement upgrading, which might include risk reduction and adaptation, allowing urban planners to engage in a complex process that is organic and dynamic (see Goethert et al. 1992; Hamdi 2010; Hamdi and Goethert 1997; Stein 2010).

- Community exchange programmes: programmes to exchange knowledge and expertise between communities at risk and between their residents and the organizations serving them.

BOX 4.4

Institutional approaches for adaptation in the global south: the cities of Cape Town and Durban

In 2006, the city of Cape Town in South Africa published a Framework for Adaptation to Climate Change in response to the potential impacts associated with climate change and increased disaster risk (Mukheibir and Ziervogel 2006). This framework calls for the development of a City Adaptation Plan of Action (CAPA), which is currently being drafted. In the city of Durban, since 1999 numerous climate change plans have been developed, as well as strategies for integrating climate change adaptation into the disaster management plan and various urban sector plans.

Emanating from the different frameworks and plans for adaptation planning and risk management, both Cape Town and Durban have initiated a number of projects in poor, high-risk settlements (mostly informal) to support communities to sustain their livelihoods in the face of increasing disaster risk. Concrete actions include involving the community in risk and vulnerability mapping, identifying adaptation options and assessing the sustainability of the various options (Carwright *et al.* 2008; Roberts 2008, 2010). This has enabled communities to better understand their own risk profile and adopt risk reduction strategies. Other projects include a focus on food and water security in which alternative crops to maize are sought and water-harvesting technologies are investigated. Furthermore, reforestation projects were started through which the community has the opportunity to generate income by collecting seeds and growing, planting and maintaining indigenous trees, while simultaneously rehabilitating ecosystems and the associated ecosystem services (Roberts 2008, 2010).

Due to the described efforts, Cape Town has been named role model city for informal settlement upgrading and ecosystems protection as part of the UNISDR 'Making Cities Resilient' Campaign, and Durban is widely regarded as a 'leader' in climate change adaptation, especially in relation to developing contexts.

By Dewald Van Niekerk and Willemien Faling

In addition to the described differences, in many low- and middle-income countries, risk reduction and adaptation are (and have to be) closely related to national and local governments' work in poverty reduction (see Table 4.6). Reducing poverty and vulnerability levels among the general population is a necessary step to improve the adaptive capacity (although poverty reduction

Figure 4.4 Environmental 'traffic lights' in Manizales, Colombia. The upper sign shows social, economic and environmental indicators for the quality of the municipality's 11 different districts (which includes their level of disaster risk). The lower illustration urges people to take action and not 'remain seated' if their district requires improvements. Source: Christine Wamsler.

does not 'by default' reduce risk). In India, government officials view anti-poverty programmes as the best adaptation strategy (Meister *et al.* 2009). In 2010/2011 more than 15 per cent of India's national budget was spent on adaptation measures, of which the majority (72 per cent, or 11 per cent of the national budget) was invested in poverty reduction (Panda 2011). The integration of risk reduction and adaptation into pro-poor planning is thus crucial. Green or ecosystem-based approaches, generally more strongly promoted in northern cities (see Section 4.1), can also be a good way to address poverty. Municipal officials of Durban in South Africa argue that the urban poor are often directly dependent on ecosystem services for their basic needs and well-being. Consequently, they advocate ecosystem-based adaptation, which is 'not simply about saving ecosystems, but rather about using ecosystems to help "save" people and the resources on which they depend' (Roberts *et al.* 2012: 172) (see also Box 4.4).

Let's turn to urban risk reduction and adaptation in high-income nations (see Tables 4.6 and 4.9). Three important aspects can be highlighted. First, northern cities are generally characterised by better social protection systems. This results in higher levels of social security in terms of income security and access to essential services, in particular health and education, which enhances people's adaptive capacity. Second, in many northern cities there is strong support for green, ecosystem-based or environmental-oriented measures. This is based on the widespread belief that so-called 'green cities' benefit the economy by leading to, for instance, increased growth, job creation, inward investment, innovation, entrepreneurship and attraction of skilled workers (Rode *et al.* 2012). Nevertheless, related measures and progress are often purely focused on climate change mitigation, and address issues such as waste and recycling, water pollution, road congestion and urban sprawl. With the growing understanding that green cities are not necessarily the best adapted ones, city authorities are increasingly searching for better synergies between climate change mitigation and adaptation (see Boxes 2.11 and 3.6). Finally, whilst in the north participatory urban risk reduction and adaptation is still in the early stages, there is increasing interest in methods for involving citizens more effectively. In the Baltic region, scenario workshops and citizen summits have for instance been used to involve citizens in prioritizing actions related to future land use and to pass on their views to the city authorities (BaltCICA 2012). Nevertheless, current measures to increase participation are generally restricted to some kind of punctual information transfer.

4.4 City-to-city (and -citizens) knowledge transfer

In any strategy that aims to help those most at risk to optimize their adaptation to changing climate and hazard conditions, knowledge transfer is a vital component (IPCC 2000). Increasing interest is thus given to horizontal knowledge transfer between cities. Vertical flows, such as city–country or city–citizens connections, however, receive less attention. The importance placed on knowledge transfer between cities relates to the fact that city authorities are generally seen as today's key players in urban risk reduction and adaptation (see Boxes 2.12 and 2.13).

The terms 'knowledge transfer', 'technology transfer' and 'leapfrogging', when used with reference to sustainable development, are often considered to mean a transfer from the northern to the southern hemisphere. Before continuing to read, ask yourself: could a knowledge transfer from southern to northern cities also provide valuable input to improve current approaches to risk reduction and adaptation? If yes, what type of knowledge could be transferred? And do you know of any examples of south–north knowledge transfer?

The international institutional infrastructure to enable the flow of ideas on risk reduction and adaptation between low-, middle- and high-income cities is not yet well developed and is a major hindrance to both research and action taking globally (Pelling 2012). The international focus on issues of climate change mitigation, together with the complex nature of adaptation, have constrained urban responses for risk reduction and adaptation beyond a group of pioneering cities. Consequently, many cities are grappling with the challenge of how to learn from these pioneering cities (UNHABITAT 2011). Related knowledge transfer from southern to northern cities hardly exists. In the context of climate change mitigation there are, however, several cases.

An example of south–north knowledge transfer for climate change mitigation is the Brazilian city of Curitiba, which implemented in 1974 the world's first Bus Rapid Transit (BRT) system, a high-efficient transportation system for cities involving a separate bus lane, high-frequency service, efficient ticket systems and bus stops with capacity for many passengers to smoothly get on and off. Marketed as a climate-friendly transport alternative (at a cost of about 5 per cent of a new subway system), similar BRT systems have been 'copied' and implemented worldwide in cities such as Cleveland (Ohio), Ottawa (Ontario), Rouen (France), Seattle (Washington) and Vancouver (British Columbia) (Byrnes 2012). Another example of the transfer of knowledge on climate-friendly inner-city transportation is shared taxis, which are used as a complement to the regular transport systems in many southern countries. Inspired by the Latin American *colectivo* (Spanish for 'shared taxi'), Texxi, for instance, is an SMS-based taxi service where passengers going in a similar direction can reserve seats in the same vehicle.[38] The fare is cheaper than in regular taxis and by avoiding travelling with empty seats, emissions are reduced. Texxi has successfully been implemented in Liverpool, UK, and there are plans for its introduction in other cities. A further example of knowledge transfer for climate change mitigation comes from Bogotá, Colombia, where every Sunday certain streets are closed and can only be used for walking and biking. Under the name *ciclovía* (Spanish for 'bike path'), the concept has spread to cities all over the American continent (including New York, Los Angeles and Chicago), promoting reduced car use and increased physical activity and social interaction among city dwellers.[39]

Conditional cash transfers constitute yet another example for successful south–north knowledge transfer. The Mexican programme *Oportunidades* (Spanish for 'possibilities') was the role model for *Opportunity NYC*, the first conditional cash transfer initiative to be implemented in the United States. This programme provides monetary incentives for low-income New Yorkers to engage in, for example, school education, healthcare and professional training, with the aim of breaking the poverty cycle (*The Economist* 2010).[40]

When it comes to risk reduction and adaptation, successful examples of south–north knowledge transfer can rarely be found. Existing knowledge transfer is often focused on the transfer of technological solutions from high- to low- and middle-income nations. Many cities are today grappling with the challenge of how to learn from other cities, which is especially true for south–north knowledge transfer. In the paper 'North–south/south–north partnerships: closing the "mutuality gap"', Johnson and Wilson (2006) explore mutuality and learning in practitioner-to-practitioner partnerships between local governments in the UK and Uganda. The benefits that were emphasized by the northern partners are very general and mainly 'soft' ones, such as greater cultural awareness, friendships and mutual understanding; however, as the authors highlight, potential benefits for the north could be many more (Johnson and Wilson 2006).

On the basis of the predominant measures, strategies and approaches taken for urban risk reduction and adaptation (described in Sections 4.1–4.3), fields of knowledge that could be transferred from southern to northern cities include:

- Forms of cooperation between city authorities, NGOs and civil society organizations
- Participatory decision-making methods used by city authorities and NGOs
- Possible users' involvement in service provision
- Anti-poverty agendas and their linkages with risk reduction and adaptation
- Community-based risk reduction and adaptation, including bottom-up mechanisms to assess risk and local capacities, and to plan, design, implement and monitor particular risk reduction and adaptation measures

- Citizens' coping and self-reliance strategies and their value for complementing institutional efforts (see Chapters 5 and 6)
- Examples of how to design measures that 'build on' local coping strategies (see Chapters 5 and 6)
- Possibilities to upscale community-based risk reduction and adaptation to municipal levels (e.g. by linking them to city-wide responses).[41]

In contrast to south–north knowledge transfer, south–south partnerships are more prominent. They are increasingly suggested to foster urban risk reduction and adaptation and to complement north–south partnerships (e.g. Taylor *et al.* 2008). There are various advantages to partnerships between low- and middle-income countries. They are generally characterized by more similar benefits, peer learning and resources, and the participating countries 'own' the process (without involvement from the 'north'). A telling example is the 'Technical Cooperation among Cities of Developing Countries Programme' of the CITYNET network, which consists mainly of Asian cities. It offers opportunities to learn from other member cities' success and failure stories and enables technical exchange on urban issues, including risk reduction, adaptation and the implementation of the Millennium Development Goals.[42] CITYNET also offers City-to-City Cooperation Awards to highlight the best examples of cooperation amongst city authorities.

When it comes to north–north knowledge transfer on risk reduction and adaptation, there is support by a range of governmental and private bodies at national and regional levels. Examples are, for instance, the European Union and its European Climate Adaptation Platform (CLIMATE-ADAPT) (EEA 2012) (see Section 6.8), Training Regions' 'Ask-Your-Urban-Neighbour' initiative (see Box 2.13) and Local Governments for Sustainability (ICLEI) (see Section 4.8). UNISDR's Resilient Cities Campaign also offers a platform for city-to-city learning, which has led to north–north and south–south interchanges (see Box 2.12). Cities and partners in the campaign have said that the main value for city authorities engaging in this campaign is to learn from others and share information.[43]

In contrast to horizontal flows between cities, vertical city–citizens connections receive little attention. As a result, urban dwellers often feel insufficiently informed and supported in their own efforts (see Chapter 5). They lack, for instance, information about the means and responsibilities they have to protect themselves, i.e. what the local community and the individual citizen can, and is expected to, do in order to adapt to and cope with increasing risk and disasters. Where initiatives such as general advice, counselling or formal incentives to take individual action exist, vulnerable citizens or population groups often are not aware of them (see Sections 5.6 and 6.2).

In comparison with risk reduction and adaptation, many citizens seem to feel more encouraged to support city authorities' efforts for climate change mitigation. This is at least a common feature in many high-income nations where city authorities, together with the private sector (including insurance companies),[44] actively encourage citizens to take small-scale actions to minimize greenhouse gas emissions in everyday life. Much attention is here given to how to act, live, eat and consume in a climate-smart fashion. An example is Copenhagen, which is constantly investing in more and better connected cycle tracks and cycle parking facilities, and safer conditions for cyclists. As a result, Copenhagen often appears first in rankings of bike-friendly cities due to its high share of people who cycle to work or school.[45] Another city that has managed to encourage its citizens to take actions for climate change mitigation is York in the UK. The city introduced 'The York Green Neighbourhood Challenge' with the goal of reducing CO_2 emissions by 10 per cent during 2010 (denoted 'the 10:10 target'). The campaign has encouraged different neighbourhoods to compete against each other and succeeded in reducing the average yearly carbon footprint of participants by 11 per cent. It has raised public awareness about low carbon lifestyles and has also improved community cohesion. The 'CoolRoofs' project of New York City in the United States is another example that shows that it is possible to encourage and incentivize citizens to take actions for climate change mitigation and, at the same time, support risk reduction and adaptation. In this project, the Department of Building, the City Service and residents collaborated to cover New York rooftops in a white reflective layer to improve heat insulation and thus reduce energy used for cooling (see Section 6.3.2, Figure 6.2). Coating all eligible dark rooftops in New York City could result in up to a one-degree reduction of the city's air temperature (NYCDOB 2011).

There are further examples of cities that have long histories of incentivizing their inhabitants to take local actions for sustainable urban development – actions that today could be capitalized on for risk reduction and adaptation. An example is urban agriculture, which is increasingly discussed as a means for urban risk reduction and adaptation in order to regulate local temperatures, mitigate floods and reduce urban–rural dependencies. Allotment gardens have been established since the Industrial Revolution in many cities in Europe and beyond.[46] In Sweden, allotments were, for instance, introduced around 1900 and were included in the municipal plans of Malmö and

Stockholm. Small pieces of land were rented out to working-class families to avoid the industrialization of certain unexploited areas and to promote urban gardening and agriculture, with the aim of providing recreation for citizens, improving households' economy and nutrition status and alleviating the effects of the unhealthy and overcrowded urban environments.[47] In twentieth-century Europe, allotments were invaluable when food flows were cut in times of financial crisis and war. In 1917, around 870,000 kg of potatoes were harvested from Stockholm's allotments. Likewise, on the isolated British Isles, the number of allotments grew rapidly in the food crisis following the Second World War (Schimanski 2008). Nowadays, allotments are mostly used for recreation, but still serve to conserve and promote the care of green areas in cities, and boost people's well-being, social interaction and, possibly, risk reduction and adaptation.[48]

4.5 Summary – for action taking

Existing national, regional and municipal planning systems vary greatly in terms of their priorities, the scope and extent of their powers, their regulatory tools, planning criteria and the resources with which they work. City authorities' and planners' capacity to perform and deliver thus varies from place to place and from time to time, and is today further challenged by increasing disasters, climate change and rapid urbanization (see Table 4.10). Rapidly growing cities and constantly changing risk patterns lead to city authorities not being able to supply adequate assistance (in form of housing, infrastructure and basic services), making planning for risk reduction and adaptation challenging. Urban growth might even lead to cities growing together, without effectively integrating their risk governance functions, resulting in confusion and the inability to coordinate any risk reduction and adaptation efforts.

Whilst city authorities and planners employ a variety of different measures and strategies to foster urban risk reduction and adaptation (see Tables 4.1–4.5 and 4.7–4.8), they are not, as yet, tapping into their full potential. Little attention is paid to assessing the root causes of risk, which lie deeply embedded in fragile societal conditions and the associated city–disaster linkages that increasingly make cities into hotspots of risk (see Chapter 3).

Consequently, the separate implementation of some isolated stand-alone measures for risk reduction and adaptation still seems to be more appealing to city authorities and planners than the consideration of risk and climate change in their day-to day operations. Furthermore, the grey infrastructure approach is predominant, with a focus on measures that reduce physical vulnerabilities. The green (and blue) infrastructure approach is, however, receiving increasing interest and is, especially in northern cities, widely promoted as a win–win or low-regret measure to deal with multiple hazards and uncertainties. Independent of the particular approach taken, strong emphasis is given to risk reduction and adaptation measures that improve cities' water management. Throughout human civilization and settlement, this issue has been the focus of human adaptation. Many city authorities' also place great emphasis on keeping urban disaster-prone areas free from further developments, although this is generally a highly politicized issue. Response preparedness generally receives little attention, and recovery preparedness even less.

As for the mainstreaming of risk reduction and adaptation at institutional levels, city authorities' efforts focus on the improvement of (a) their legal planning frameworks and (b) cooperation between different urban actors to advance existing risk governance structures. The borders between institutional and individual responsibilities are, in this context, not given major consideration. This can obstruct the creation of more functional urban risk governance systems, in which different actors' efforts complement and support (rather than hinder) each other. Other mainstreaming strategies receive comparatively little attention (e.g. strategies for reducing urban organizations' own risk or for improving science–policy integration) (see Table 2.7).

Whilst it can be argued that city authorities in high-income nations are generally better at dealing with 'conventional' weather extremes (such as riverine flooding), their systems and mechanisms associated with these extremes are becoming 'outdated' in the context of climate change (such as increased pluvial flooding or flash floods, which form part of its consequences). But how can city authorities or other urban actors assess their adaptive capacity to subsequently improve their risk reduction and adaptation planning?

The adaptive capacity of a city authority or another urban organization can be assessed, first by analysing how it assists people and communities in (a) reducing or avoiding hazard exposure, (b) reducing location-specific vulnerabilities, (c) preparing for disaster response and recovery, and (d) responding to and recovering from hazards and disasters (including climatic extremes and variability). Second, it is necessary to evaluate the organization's achievements in terms of its (level of) mainstreaming, i.e. the sustainable incorporation of risk reduction and adaptation into urban planning practice. If resilient cities are the aim, risk reduction and adaptation need to become an inherent part of urban planning, which requires the use of a set of different main-streaming strategies (see Section 2.7). The following guiding questions can be used by urban organizations to assess their adaptive capacities and related work.

1. What types of risk reduction and adaptation measures have been taken 'on the ground'?

   a) Identification of all measures implemented at programme/house-hold level

   b) Analysis of the identified measures as regards the way they:
      i) address risk (see Sections 2.2 and 4.1)
      ii) support communities (i.e. direct or indirect support through capacity development) (see Section 2.2)
      iii) involve different stakeholders (see Sections 2.2 and 4.3)
      iv) consider the city–disasters nexus (see Sections 3.1–3.5)
      v) combine physical with non-physical measures (e.g. combined grey and green infrastructure approach) (see Sections 3.1–3.5 and 4.1)
      vi) consider local adaptive capacities (see Sections 2.1 and 5.1–5.5).

2. How have the identified measures been realized? And what types of strategies have been taken at institutional levels to support sustainable risk reduction and adaptation planning 'on the ground'?

   a) Identification of all actions taken at institutional levels

   b) Categorization and analysis of actions at institutional levels in terms of mainstreaming strategies taken (see Boxes 2.9 and 2.10 and Sections 4.2–4.3)

   c) Categorization and analysis of the measures at programme level in terms of the mainstreaming strategies taken (see Box 2.8 and Sections 4.2–4.3).

Such an assessment provides the knowledge base required to improve urban organizations' adaptive capacity. The temporary creation of specialized departments or focal points, and the establishment of city-to-city networks (for north–south, south–north, north–north or south–south knowledge transfer) and other decision-support systems (such as online adaptation portals) can help to push the mainstreaming process further by improving access to relevant knowledge and providing required resources in terms of funding, know-how and staff.

4.6 Test yourself – or others

The following questions can be used to test yourself or others. The answer to most questions can be found explicitly in Chapter 4.

1. What are the aspects that have to be taken into consideration when analysing current planning practice for risk reduction and adaptation?
2. What is a grey infrastructure approach? Please give concrete examples of grey measures for vulnerability reduction in relation to (a) floods, (b) windstorms and (c) heatwaves.
3. What is a green infrastructure approach? Please give examples of green measures for vulnerability reduction in relation to (a) floods, (b) windstorms and (c) heatwaves.
4. What types of physical (or grey) measures exist to support preparedness for response?
5. What does the concept of 'climate proofing' entail?
6. What are win–win or no-regret measures for risk reduction and adaptation? Please provide illustrative examples.
7. What measures for risk reduction and adaptation are commonly taken by city authorities?
8. What strategies are commonly taken by a city authority to institutionalize risk reduction and adaptation (i.e. to integrate risk reduction and adaptation at institutional level)?
9. Under what circumstances should an organization engage in internal mainstreaming? And what types of measures/actions can be taken?
10. Name some of the constraints city authorities face when planning for risk reduction and adaptation? Provide concrete examples of organizational and interorganizational constraints, and discuss how the concept of mainstreaming takes them into account.
11. Why is mainstreaming in infrastructure planning especially crucial?
12. What could northern cities learn from southern cities' approaches to reduce and adapt to increasing risk? Please provide some illustrative examples.

Case studies might also be a good basis for assessing the knowledge presented in this chapter.

Test yourself scenario 4.1: sector-specific mainstreaming of risk reduction and adaptation. You are working for an international organization and are responsible for the overall coordination of its work on urban development. After a series of climate-related disasters in San Salvador, your boss calls for a meeting. During the meeting, s/he informs you about funding that your organization has received for long-term recovery support for risk reduction and adaptation in San Salvador. S/he further provides you with a list of projects run by organizations that are already working in San Salvador. In order to be able to decide what type of assistance could best complement these efforts, s/he asks you to prepare a comprehensive list of the different types of urban sectors that potentially exist, and which could be supported as regards the mainstreaming of risk reduction and adaptation.

Task. Discuss or answer the following issues/questions:

1. What are the urban sectors that need to be considered when main-streaming risk reduction and adaptation into urban planning?
2. Select a specific urban sector and provide examples for potential measures (including at least one physical, environmental, social, economic and political measure).
3. Discuss the generic issues within or across sectors that can be planned for to prepare for response and recovery.

Test yourself scenario 4.2: south–north knowledge transfer for risk reduction and adaptation planning. Knowledge transfer is often understood as the transfer of science and know-how from the northern to the southern hemisphere (i.e. from high-income nations to low- and middle-income nations). Through answering the questions below, you are asked to investigate the potential advantage of knowledge also spreading from south to north.

Task. Discuss or answer the following issues/questions:

1. Review the main differences between current planning practice for risk reduction and adaptation in northern and southern cities (type of measures, strategies, key actors, etc.).
2. Take a look at the challenges and constraints city authorities and planners can face (see Table 4.10). Do you think that some of these issues are more frequent in southern cities? Why?
3. Conversely, what do you think are the respective advantages in each context?
4. Finally, give examples of ways in which knowledge on adaptation and risk reduction has spread, or could beneficially be spread, from south to north.

Test yourself scenario 4.3: analysing institutional approaches for risk reduction and adaptation. This scenario is the analysis of the development of Augustenborg district in Malmö, Sweden. Built in the 1950s,

Augustenborg was initially considered a highly successful urban development project. By the 1970s, the neighbourhood had, however, fallen into
decline, with problems such as damp, poor external appearance and annual
flooding. As people moved out, many flats remained empty and the tenants
who stayed suffered increasingly from a high level of unemployment and
growing marginalization. In 1998, a regeneration project, known as the
Eco-City Augustenborg (*Ekostaden Augustenborg*), was initiated in partnership between Malmö City, the MKB Housing Company, the housing
landlord and local residents. Project measures included a new stormwater
management system in which water from rooftops and other impervious
surfaces is collected and channelled through canals, ditches, ponds and
wetlands before finally entering a traditional closed subsurface stormwater
system. Furthermore, on all new constructions in the neighbourhood, as
well as on some of the old buildings, green roofs were installed. These
intercept on average half of the total rainwater run-off and have a significant cooling effect in the summer. Another measure was the improvement
of green spaces so that they can temporarily be flooded, thus slowing the
entry of rainwater into the stormwater system. During normal times, they
provide a venue for residents' leisure activities, small-scale food production
and children's play. Measures to reduce CO_2 emissions were also taken,
notably by installing solar thermal panels and a ground source heat pump
system to produce heat, photovoltaic panels to generate electricity, better
insulation of pipework and facades to avoid heat loss, a reduction in the
speed limit on so-called 'garden roads' (also to encourage walking and
cycling), a community car pool and a recycling scheme aiming to recycle
90 per cent of all waste. Throughout the development of Eco-City Augustenborg, residents and people working in the area have been involved in the
project. Local knowledge and capacities were considered. Some local residents got even more involved in the project. One such, an amateur water
enthusiast, provided the city authority with knowledge and ideas on how to
design the stormwater system. In addition, the maintenance contractors
offered jobs to local unemployed young adults. Today, Eco-City Augustenborg
is an attractive place to live. The neighbourhood's improved disaster resilience was put to the test during a major flood in 2007. Augustenborg coped
very well, and much better than nearby districts. In addition, Augustenborg's green roofs feature the world's first botanical roof garden. Covering
more than 9,000 m^2, it works as a demonstration and research facility, and
students from different disciplines and various universities regularly visit
for educational reasons. For more information on the project, see http://
www.malmo.se/English/Sustainable-City-Development/Augustenborg-Eco-
City.html

Task. Discuss or answer the following issues/questions:

1. Why do you think the regeneration project Eco-City Augustenborg has been so successful?
2. Analyse the project by identifying (a) the employed risk reduction and adaptation measures and (b) the mainstreaming strategies taken.
3. Analyse the measures taken for both climate change mitigation and climate change adaptation as regards their link between the urban fabric and its ecological, social, economic, institutional and/or political features.

4.7 Guide to further reading

Bicknell, J., Dodman, D., and Satterthwaite, D., 2009. *Adapting cities to climate change: understanding and addressing the development challenges*. London: Earthscan, 3–47.

Carmin, J., Nadkarni, N., and Rhie, C., 2012. *Progress and challenges in urban climate adaptation planning: results of a global survey*. Cambridge, MA: MIT.

CoR (Committee of the Regions), eds., 2011. *Adaptation to climate change: policy instruments for adaptation to climate change in big European cities and metropolitan areas*. Brussels: European Union, Committee of the Regions.

Davoudi, S., Crawford, J., and Mehmood, A., eds., 2009. *Planning for climate change: strategies for mitigation and adaptation for spatial planners*. Abingdon and New York: Routledge.

EEA (European Environment Agency), 2012. *Urban adaptation to climate change in Europe: challenges and opportunities for cities together with supportive national and European policies* (No. 2). Copenhagen: EEA.

Kazmierczak, A. and Carter, J., 2010. *Adaptation to climate change using green and blue infrastructure: a database of case studies*. Manchester, UK: University of Manchester.

Satterthwaite, D., 2011a. *What role for low-income communities in urban areas in disaster risk reduction?* Background paper prepared for the 'Global Assessment Report on Disaster Risk Reduction (GAR 2011)'. Geneva: UNISDR.

Satterthwaite, D., *et al.*, 2007b. *Adapting to climate change in urban areas: the possibilities and constraints in low- and middle-income nations*. Human Settlements Discussion Paper Series. Theme: Climate change and cities (1). London: IIED.

Shah, F., and Ranghieri, F. 2012. *A workbook on planning for urban resilience in the face of disasters: adapting experiences from Vietnam's cities to other cities*. Washington, DC: The World Bank.

UNISDR (United Nations Office for Disaster Risk Reduction), 2011a. *Climate change adaptation and disaster risk reduction in Europe: a review of risk governance*. UNISDR, EUR-OPA, Council of Europe.

UNISDR (United Nations Office for Disaster Risk Reduction), 2012c. *Making Cities Resilient report 2012: my city is getting ready! A global snapshot of how local governments reduce disaster risk*. Geneva: UNISDR.

See also recommended readings included at the end of the other book chapters.

4.8 Web resources

Asian Cities Climate Change Resilience Network (ACCCRN). The ACCCRN aims to catalyse attention, funding and action on building climate change resilience for poor and vulnerable people by creating robust models and methodologies for assessing and addressing risk through active engagement and analysis of various cities: http://www.acccrn.org

CITYNET. CITYNET is a regional network of urban stakeholders for the Asia–Pacific region. It assists cities and local governments to provide better services to citizens: http://www.citynet-ap.org

Earthquake and Megacities Initiative (EMI). EMI is an international, non-profit scientific organization dedicated to the reduction of disaster risk in megacities and major metropolises: http://www.emi-megacities.org

ICLEI Local Governments for Sustainability. ICLEI is an international association of over 1,220 local government members who are committed to sustainable development, as well as national and regional local government organizations that have made a commitment to sustainable development. ICLEI serves as Secretariat of the World Mayors Council on Climate Change. In 2010, ICLEI and the city of Bonn, Germany launched Resilient Cities, the first World Congress on Cities and Adaptation (in 2012 renamed as Global Forum on Urban Resilience and Adaptation). ICLEI is also a network of networks including the Resilient Cities Network, the EcoCities Network and the GreenClimateCities Network: http://www.iclei.org

ResilientCity.org. ResilientCity.org is an open, not-for-profit network of urban planners, architects, designers, engineers and landscape architects whose mission is to develop creative, practical and implementable planning and design strategies that help increase our communities' and cities' resilience to the future shocks and stresses associated with climate change, environmental degradation and resource shortages, in the context of global population growth: http://www.ResilientCity.org

United Cities and Local Governments (UCLG). UCLG is the world's largest organization of local and regional governments working in 140 countries to promote democratic local self-government, disaster awareness campaigns and putting disaster risk reduction on the local and regional political agenda: http://www.cities-local-governments.org

United Nations Framework Convention on Climate Change (UNFCCC) Database on Local Coping Strategies. The database is aimed at facilitating the transfer of local knowledge and experience from institutions and/or communities that have engaged in adaptation to specific hazards or climatic conditions to institutions and/or communities that may just be starting to experience such conditions due to climate change: http://maindb.unfccc.int/public/adaptation

Urban Climate Change Research Network (UCCRN). The UCCRN is a consortium of experts from academic and research institutions around the world established to facilitate and build connections between science, policy and practice regarding effective climate change mitigation and adaptation. It is designed to enhance cutting-edge scientific, economic and planning-related research and to promote knowledge sharing among researchers and urban decision-makers about all aspects of climate change and cities: http://uccrn.org

URBAN-NEXUS. URBAN-NEXUS is a coordination and support action funded by the European Union under the Seventh Framework Programme, with the aim of enabling knowledge transfer and dialogue, and forming long-lasting partnerships in and between cities and regions to deal with integrated sustainable urban development: http://www.urban-nexus.eu

World Urban Campaign (WUC). WUC is a global coalition of public, private and civil society partners united by the common desire to advocate on the positive role of cities around the world, and to promote sustainable urbanization policies, strategies and practices. Launched in Rio de Janeiro at the Fifth Session of the World Urban Forum in March 2010, the campaign is coordinated by UNHABITAT and governed by a Steering Committee of partners: http://www.worldurbancampaign.org

See also web resources included at the end of the other book chapters.

5 City dwellers' own ways to reduce and adapt to urban risk

Learning objectives

- To obtain a systematic overview of citizens' own efforts to reduce and adapt to urban risk
- To reflect upon the strengths and weaknesses of people's coping capacities, especially in a context of climate change
- To understand the importance of taking into account people's coping capacities for achieving sustainable risk reduction and adaptation

Humans have always adapted to environmental changes, including variations in weather and hazards.[1] This accumulated local capacity to adapt is increasingly recognized to be critical in reducing risk and vulnerability (e.g. Banks *et al.* 2011; Dodman and Mitlin 2011; Eriksen *et al.* 2011; McAdoo *et al.* 2009; Shaw *et al.* 2008; Soltesova *et al.* 2012; see Wamsler 2007a). However, city dwellers' adaptive capacities are little known, poorly documented and hardly considered in city authorities' work. Furthermore, the environmental changes humanity is facing today are happening at a previously unseen rate and magnitude, and are deeply intertwined with other complex processes such as urbanization and globalization. This inevitably places new demands on people's adaptive capacities.

Citizens' own ways to reduce and adapt to urban risk are often called coping strategies, private adaptation, autonomous adaptation or adaptive behaviour (see Section 2.1).[2] *Used* adaptive capacity is seen in how citizens reduce and adapt to urban risk. People take measures during 'normal' times with the aim of being less affected by potential small-scale or exceptionally large-scale hazard impacts. In addition, during and after hazard impacts, citizens employ reactive *ad hoc* measures to respond and recover, which they have not prepared beforehand.[3]

> We are always trying to improve, little by little, step by step, in order to
> become more secure.
>
> > (San Salvador resident)

This statement by a resident of San Salvador, the capital of El Salvador, illustrates the constant efforts some city dwellers put into coping with disasters
and other climate change impacts. A quite different statement comes from a
resident of Helsingborg City in Sweden who was affected by floods in 2011:

> There is nothing I do. It is not possible to protect myself [from future
> disaster impacts]. Nature is what it is.
>
> > (Helsingborg resident)[4]

These two quotations illustrate the wide spectrum of citizens' levels of
active engagement in risk reduction and adaptation. Before continuing to
read, ask yourself: are there any concrete measures that citizens living in
urban areas exposed to hazards and climate change take to reduce or adapt
to associated disasters and risk? And how could we analyse such measures
in a systematic way in order to gain an understanding of people's adaptive
capacities?

People's coping strategies can first be analysed, based on their respective
objectives, into hazard reduction and avoidance, vulnerability reduction,
response preparedness and *ad hoc* responses, recovery preparedness and
ad hoc recovery (see Box 2.4). Second, citizens' measures can be assessed
in relation to their thematic foci, being physical,[5] environmental, socio-
cultural, economic, political or institutional. In addition, they can be
reviewed on the basis of their underlying patterns of social behaviour, either
individualistic, communitarian, hierarchical or fatalistic.[6] 'Individualistic'
behaviour can be characterized as self-help, fixing things oneself (without
assistance from people outside one's own household). 'Communitarian'
behaviour is based on the belief that everybody sinks or swims together
and is characterized by community efforts. 'Hierarchical' patterns relate to
the belief in, and reliance on, authority structures and strong leadership for
assistance, control and organization. 'Fatalistic' behaviour is a 'non-strategy'
for survival. It is based on the idea that taking action, or not, has the same
(negative) result.[7] Finally, it can further be assessed whether citizens' measures were (a) taken well in advance or shortly before potential hazard
impacts, (b) planned or *ad hoc*, (c) intentional/deliberate or unintentional
(i.e. taken with or without intent to reduce or adapt to risk), and (d) whether
they were supported by any governmental authorities or aid organizations.[8]
Taking into account these aspects, do you think that the measures of city

dwellers who face the same types of hazards are similar worldwide? And do you think that there are urban–rural differences?

Unfortunately, there is little knowledge on citizens' own efforts and their potential capacities to reduce and adapt to increasing risk and disasters, which makes related analyses very challenging. This is associated with the fact that most of the current work on risk reduction and adaptation is on macro-scale issues (Morss *et al.* 2011). In addition, existing research on local coping strategies focuses mainly on rural areas, and there are hardly any analyses that contrast rural with urban dwellers' strategies. But what kind of measures can be found?

The following sections provide an overview of existing risk reduction and adaptation measures, including planned, *ad hoc*, intentional and unintentional measures taken in the pre- and post-disaster context. They are based on the analysis and comparison of primary data, single-case and cross-case studies.[9]

5.1 Coping through hazard reduction and avoidance

> I got old car tyres from my neighbours' repair shop to build a floodwall next to the river. But if the situation gets worse, I consider moving away.
> (See Wamsler 2007a; adapted from real situation)[10]

> I ran away with my children and got separated from my husband, because he never wanted to leave this place.
> (Wamsler *et al.* 2012)

To reduce their hazard exposure, urban dwellers around the world use a variety of measures. Table 5.1 provides an overview of the measures taken for hazard reduction and avoidance.

Most actions taken for hazard reduction and avoidance are so-called grey or green measures that are focused on improving physical or environmental conditions. City dwellers build, for instance, small embankments to improve protection from river or sea flooding, and construct retaining walls to protect against landslides. Materials and construction techniques vary considerably and include bricks, cement, stones, nylon bags filled with soil and cement, compacted soil, old tyres or a combination of old tyres, stones, cement and soil. In some places, such as Medellín in Colombia, residents also use biodegradable sandbags filled with soil and seeds, which transform over time into green retaining walls when the sandbags burst under pressure from the growing plants (see Figure 4.3).[11] Other people grow certain plants

Table 5.1 Coping strategies for hazard reduction and avoidance. This categorization mainly includes individually taken physical (grey) and environmental (green) measures[a]

Hazard	Illustrative examples of activities/measures
Precipitation, flood and sea-level rise	Construction of small levees, dams and embankments to protect the settlement from flood
	Landfill/expansion to reduce erosion (e.g. next to the sea or a river)
	Planting (or 'fighting' deforestation) to preserve natural flood protection (in surrounding areas of urban settlements/cities)
Landslides and erosion	Construction of retaining walls (e.g. made of old car tyres or pre-seeded biodegradable sandbags)
	Compacting soil on slopes
	Planting vegetation to prevent landslides
	Use of pre-seeded bio-mats, which quickly re-grass steep slopes and reduce erosion (commercial product)
	Putting plastic sheets on slopes
	Reduction of uncontrolled water flows that cause landslides (see also under vulnerability reduction, Section 5.2)
Temperature, heat- and coldwaves	Measures to reduce the heat island effect (see also under vulnerability reduction, Section 5.2)
Drought and water scarcity	Measures to increase the recharge of dug wells (open or closed, hand-drawn or equipped with hand pumps) such as small dams or bunds
Fires	Improvements to electrical connections in the house that may cause fires
Wind- and snowstorm	Planting to create windbreaks in surrounding areas of urban settlements/cities
Earthquake and volcanic eruptions	(No hazard reduction for earthquakes or volcanic eruptions)
Multi-hazard measures[b]	Avoiding extending their homes into hazard-prone location
	Moving permanently to a safer house/location
	Construction of fences to prevent children from getting close to risk zones
	Moving into a risk area in order to be included in post-disaster assistance (e.g. resettlement projects)
	See also Table 5.3 under 'Psychological and/or emotional support mechanisms'

Notes:

[a] The categorization of measures is not always clear-cut and there are no doubt overlaps (see Tables 5.1–5.6).

[b] All types of measures for climate mitigation can be considered to be part of hazard reduction; however, these have not been specifically analysed for or included in this table. Since there are not many non-hazard specific measures, both hazard-specific and non-hazard specific measures are included in this table.

to stabilize the soil, to counteract soil erosion and, in some instances, to create windbreaks. Covering slopes with plastic sheets to prevent landslides is another measure that is common in many southern cities.

City dwellers further take physical (grey) and environmental (green) measures that reduce both urban hazards and, at the same time, location-specific vulnerabilities. Examples are the reduction of uncontrolled water flows that can be the (sole) cause of landslides, improvements to electric connections that may cause fires, and measures to reduce the heat island effect, thus also reducing people's vulnerability to increasing climate-related temperatures.

Furthermore, citizens deploy measures to avoid hazards, for instance by not extending their homes into known hazard-prone locations, moving to a safer house or location, or imposing restrictions such as putting up fences to prevent children from getting close to risk zones. A diametrically opposite strategy is to move into a risk area with the aim of being included in post-disaster resettlement programmes, i.e. by first increasing one's hazard exposure.

Most of the measures for hazard reduction and avoidance are taken at an individual level, but there are also examples of community-based action. Residents, for instance, join forces to obtain construction materials or to build simple flood or retaining walls, commonly seen in marginal settlements that are at risk from floods or landslides.

5.2 Coping through vulnerability reduction

> Since we have changed the direction of the roof inclination, I have fewer problems with water infiltration and landslides.
> (See Wamsler 2007a; adapted from real situation)

> . . . people have learned the hard way. If you go around the compound now, people are using burnt bricks or cement blocks because houses built from these are stronger.
> (Simatele 2010: 21)

> Our furniture has been custom-made to help keep our things dry from the water.
> (Douglas et al. 2008: 197)

Most deliberate risk reduction and adaptation measures aim to reduce physical vulnerability (Table 5.2). Measures taken to increase structural protection from rain, floods, landslides, windstorms, earthquakes, cold and heat are, to a certain extent, already taken into account during the planning, construction or maintenance of many buildings. Important factors include the depth of foundations, the length of roof projections, the use of shutters, the height of doorsills and regular house painting to prevent water infiltration. Many urban dwellers are, however, increasingly forced to make even greater efforts to improve and adapt their living conditions. They replace doors with more flood-proof ones, increase the inclination of roofs for better rainwater run-off without damaging roof constructions, change the directions of roof inclinations to discharge run-off water without causing erosion, improve wall and roof insulation, close cellar windows in flood areas, or construct additional drains (Table 5.2). With respect to the last, there are cases in both the northern and southern hemisphere where citizens have illegally connected these drains to adjacent sewers. To protect themselves from heat, people install additional shadow devices or build extra rooms in cellars. Another physical measure comes from Medellín, Colombia, where residents protect their houses from flooding and water infiltration with long projecting pipes or rain gutters, so-called 'urban showers', in order to discharge rainwater towards the streets (Figure 5.1).[12]

Innovative examples of how citizens reduce their physical vulnerability with structural improvements can be found for nearly all types of hazards, especially in areas with little support from governmental authorities. Examples include detachable roofs which can be removed to create firebreaks, floating houses, the use of wooden plank flooring and the creation of outlets at the rear of the house which allow the water level to sink faster during flooding,

Figure 5.1 'Urban showers' in Medellín, Colombia. Source: www.inteligenciascolectivas.org

and the construction of housing on poles (see Figure 2.3) or other techniques that allow the increase in the height of the floor to mitigate floods and improve ventilation.[13] In earthquake-prone areas in Nigeria, people build houses with a special technique that uses old plastic bottles. The result is said to be 20 times stronger than a brick construction, cost-efficient and environmentally friendly.[14]

In addition to these structural improvements to houses, citizens also employ other types of physical measures to reduce their location-specific vulnerability. Examples include putting wood or bricks on the roof to secure and hold it in place during windstorms, increasing the height of furniture in flood-prone areas or gluing objects on to furniture to prevent them falling during earthquakes. Related commercialized measures used are special non-slip mats and metal brackets to prevent objects and furniture from falling.[15] To mitigate heat, citizens use curtains or roller blinds made of bright material, paint their houses white, which can lower indoor temperatures up to 30 per cent (Arrieta 2012; Ben Cheikh and Bouchair 2008), or create ventilation holes that are covered with various materials (such as empty cement bags), which also provides protection against mosquitoes (Table 5.2). In some cases, common rural practices for vulnerability reduction have been 'brought' to the urban setting. An example is city dwellers' use of canopies (made of cloth and hung under the ceiling) to reduce heat in Dhaka, Bangladesh, which is a traditional and common measure in rural areas (Jabeen *et al.* 2010).

Another physical or socio-physical measure is the exchange of rooms, with more vulnerable elderly or disabled people inhabiting the less dangerous ones (Table 5.3). This addresses the link between disasters and the urban fabric's physical and socio-economic features (see Section 3.1). The following statement from a San Salvador resident illustrates the importance of such measures:

> I have always been living on the river banks, but when I could still walk this did not worry or afflict me. But now it does.
>
> (Wamsler *et al.* 2012)

Physical measures are further often combined with environmental (green) measures. To mitigate heat, residents use potted plants and grow creeping plants (such as passion fruit vines) to cover walls and roofing. Trees and other vegetation are also used as a natural protection against landslides. Trees can, however, cause complex problems in densely populated settlements because their branches may injure people and damage infrastructure in

Table 5.2 Coping strategies for vulnerability reduction (hazard-specific). This categorization mainly includes individually taken physical (grey) and environmental (green) measures. [a]

Hazard	Illustrative examples of activities/measures
Precipitation, flood and sea-level rise	Construction or improvement of drainage systems, including the (illegal) connection of private drains to the public system
	Taking up paving stones in one's back garden so it can be used as a 'soak away'
	Construction of 'soak pits'
	Regularly painting the house to mitigate damage caused by water infiltration
	Raising doorsill levels or building water barriers in front of the houses
	Use of wooden plank flooring, which allows water levels to sink fast (during rainfall/flooding)
	Increasing the height of houses with plinths or stilts/poles
	Increasing the inclination of roofing, or changing the direction of roofing (so that rainwater is discharged without causing damage)
	Prolongation of roof projections or rainwater gutters/pipes, so-called 'urban showers' (so that rainwater is discharged without causing damage; see Figure 5.1)
	Improving electricity installations by covering cables, or raising electrical connections (and other service lines) to a higher level, out of reach of expected flood levels
	Construction of higher storage facilities. Increasing the height of furniture and storage facilities (e.g. by adding plinths) or having furniture custom-made to help keep possessions dry during floods (e.g. unusually high tables and wardrobes on which people can sit or keep valuables)
	Replacement of mud walls with brick walls, wooden pillars with metal ones and corrugated iron with more durable materials (e.g. *duralita*) (to better withstand rain and/or floodwater)
	Construction of water outlets in houses for easy outflow of water
	Cementing streets so that children, in the case of flooding, do not sink into the mud
	Cleaning waste from slopes, avoiding littering and regularly cleaning water gutters (to mitigate flooding and related contamination)

Table 5.2 (continued)

Hazard	Illustrative examples of activities/measures
	Clearing objects that block the flow of rivers, such as tyres, plastic sheets, mattresses and branches
	Repairing public infrastructure that passes through the settlement, such as wastewater pipes (to avoid related flooding and contamination)
	Construction of floating houses that rise along with the water level
	Permanently closing cellar windows
Landslides and erosion	Construction of deeper foundations
	Cutting down branches or trees located close to houses
	Prolongation of roof projections and rainwater eaves, and changing direction of roofing so that rainwater is discharged without causing damage/landslides
	Building fences around the house (to hold back soil e.g. made of corrugated iron, mattress springs, wooden pillars and wire netting)
	Changing the location of services such as latrines and wash places in low-income settlements and filling in former latrine holes with earth, stones and/or cement
	Digging water channels or building provisional water channels with corrugated iron or cement (to discharge rainwater without causing damage/landslides)
	Compacting soil and strengthening pathways by covering them with (additional) cement and filling in cracks
	Replacing eroded earth with new earth
	Putting plastic sheets on slopes, even during the entire year
Temperature, heat- and coldwaves	Construction of houses with openings to improve ventilation (e.g. so that air can flow through the house horizontally or vertically through convection)
	Insulating room ceilings with different materials to reduce indoor heat
	Painting walls and/or roofs white to reduce indoor heat
	Growing creepers (leafy vines) to cover walls and (corrugated iron) roofs to reduce indoor heat
	Planting vegetation to reduce heat and/or 'fighting' deforestation
	Installation of shutters or other shading devices in front of windows
	Use of reflective curtains or roller blinds

(continued)

Table 5.2 (continued)

Hazard	Illustrative examples of activities/measures
Drought, water scarcity	Reduction of water use (e.g. recycling of grey water by using water from laundry for flushing the toilet, installation of water-saving devices, or rainwater harvesting such as water butts to collect rainwater for domestic use or systems that collect rainwater from roofs)
	Construction of additional wells
	Switching to vegetarian diet (because meat production requires large amounts of water and energy in comparison with grains and vegetables)
Fire	Use of fireproof material
	Detachable roofs, which can be removed to create a firebreak
Wind- and snowstorm	Cutting down bigger branches and trees located close to houses
	Improvement of roof fixing
	Putting objects on the roof to hold it in place during windstorms
	Strapping houses to the ground
	Planting to create windbreaks (within settlements)
Earthquake and volcanic eruptions	Gluing objects on to furniture and/or furniture on to floor
	Improving roof constructions (fixing and reduction of weight to better withstand earthquakes)
	Replacement of mud walls with brick walls, wooden pillars with metal ones and corrugated iron with more durable materials
	Switching from building with, e.g., bricks to a construction technique using recycled plastic bottles
	Replacement or strengthening of wooden pillars and beams
	Cutting down bigger branches and trees located close to houses
	Compacting soil and strengthening pathways by covering them with (additional) cement and filling in cracks (to minimize damage caused by earthquakes)
	Filling in former latrine holes with earth, stones and/or cement
	Close to volcanoes, increasing roof inclinations to avoid overload caused by volcanic ash and use of construction material that does not oxidize

Note:
ᵃ The categorization of measures is not always clear-cut and there are no doubt overlaps (see Tables 5.1–5.6).

adverse weather. Consequently, they are often cut down, rather than pruned, to prevent falling branches causing damage.

Furthermore, city dwellers combine physical (grey) and environmental (green) measures to mitigate floods, landslides and water scarcity. For example, in flood areas people construct soak pits to allow rainwater to filter into the ground instead of forming puddles,[16] or take up paving stones in their back yards to create a 'soak away'. The latter was one of the few measures employed by residents in Manchester, UK in answer to the 2004 and 2006 floods (Wamsler and Lawson 2012). People also draw upon local knowledge on the natural environment to decide where and how best to construct their houses. Water scarcity is dealt with by a range of different measures that reduce water use. Rainwater is for instance channelled off roofs into small containers for household use and larger tanks for communal use (Ayers and Forsyth 2009). City dwellers also collect rainwater in plastic tanks and then distribute it with a drip irrigation system for watering plants (i.e. water hoses with small holes).[17] Examples of commercialized measures to cope with water scarcity, control soil erosion and strengthen slopes are water-saving devices put into toilets' water cisterns to reduce the amount used for every flush,[18] or pre-seeded bio-mats that allow for the speedy re-grassing of steep slopes.[19]

Apart from these physical and environmental measures, urban residents also deploy economic measures for vulnerability reduction, which are aimed at increasing household income and income security (Table 5.3). Many examples of deliberate measures can be found in low- and middle-income countries. They include taking low-risk jobs or jobs with differing risk profiles, economic diversification at individual and household level to reduce dependence on specific income sources, and taking jobs outside the neighbourhood to reduce the impact of local disasters (this last is an example of a measure that explicitly addresses the link between disasters, the urban fabric and the urban economy; see Section 3.1). Economic diversification is vital given the impact of disasters on people's livelihoods – sometimes even when they do not occur, as illustrated by a resident of Caye Caulker, Belize:

> It's not whether it hits or it doesn't. As long as it is expected to hit anywhere around, the tourists are immediately going to avoid this area for a good two to three weeks, even if the hurricane has passed. And that does a huge amount of damage to your revenue and business total. A hurricane affects whether it hits or it doesn't.
>
> (Esdahl 2011: 14)

Table 5.3 Coping strategies for vulnerability reduction (non-hazard specific). This categorization mainly includes social, economic and institutional measures. Individualistic and hierarchical measures are more common than communitarian ones.[a]

Strategy	Illustrative examples of activities/measures
Increasing household income and income security	Income diversification at individual or household level
	Taking jobs outside risk areas
	Taking on income activities with differing risk profiles (e.g. in different geographic areas)
	Taking on low-risk activities (i.e. jobs unaffected by local disasters or jobs that might even profit from disasters)
(Improving) access to formal assistance and information	'Fighting' for legal tenure in order to gain access to formal assistance, legal protection and other services (see also under preparedness for recovery)
	Moving into a risk area in order to be included in planned programmes for risk reduction (e.g. of city authorities or NGOs)
	Asking for assistance to reduce vulnerability (e.g. from political parties, faith-based groups and/or the municipality)
	Looking for risk information and related measures (e.g. from city authorities, the internet)
	Having engineers look at the house to evaluate physical risk and identify potential solutions
Creating reciprocal family networks	Creation of extensive reciprocal and dependent relationships (e.g. mutual support through labour and/or income of family members, such as regular 'income' through remittances)
	Encouragement of dependents and other family members to achieve improved economic and/or educational status
Creating social cohesion, solidarity and reciprocal relationships	Knowing well and interacting with people from the neighbourhood (e.g. buying from local shops, offering labour when needed, employing community members for small jobs)
	Engaging in community matters and community-based decision-making
	Sharing resources with neighbours (e.g. sharing tools and household equipment, etc.)
	Learning from friends and neighbours
	Exchange of political votes for short-term assistance from political parties, such as food, (money for) retaining walls, etc.

Table 5.3 (continued)

Strategy	Illustrative examples of activities/measures
Creating organizational structures for risk reduction and adaptation	Establishment of local committees for risk reduction
	Lobbying for the inclusion of risk reduction activities in the work portfolio of the local executive committee
	Sending children to schools outside own settlement (where the education is less disturbed by disasters, poor infrastructure, violence, etc.)
Psychological and/or emotional support mechanisms	Reliance on a hierarchical system to supply assistance for risk reduction and adaptation
	Full reliance on support from family members
	Accepting and/or downplaying the existing level of risk
	Seeking emotional support with social networks, such as family, relatives, neighbours, spiritual and religious groups
	Religious, cultural and traditional behaviour/beliefs (such as having an onion with you in a bag to avoid heat stress [India])
Physical multi-hazard measures	Improvement in electrical connections in the house to prevent electricity-related accidents/fire as secondary hazard
	Changing rooms/apartments so that more vulnerable people obtain the ones at lower risk
	All types of house improvements which lead to increased well-being and the prevention of illnesses (e.g. reduced dampness)
	Use of traditional or indigenous knowledge to build adequately and/or in lower-risk areas (e.g. knowledge on storm routes, wind patterns, local rain corridors or being able to tell the height of the water table based on the height of birds' nests near rivers or the presence of certain plant species)

Note:
a The categorization of measures is not always clear-cut and there are no doubt overlaps (see Tables 5.1–5.6).

Economic diversification means that urban dwellers and their families undertake many different income-generating activities. Renting out rooms, running a home-based business and having various service-sector jobs are common income sources, which the urban poor often 'run' in parallel.[20] Economic diversification not only makes urban dwellers less vulnerable, it can also enable them to recover (more) quickly from hazard impacts (see Table 5.6).

More socially oriented measures are related to the creation of social cohesion, solidarity and reciprocal relationships with family members, neighbours and other community members (Table 5.3). The case of the 2003 European heatwave cruelly illustrates the importance of social cohesion and networks. Many victims were elderly people who lived alone in neighbourhoods that lacked a sense of community. They were trapped inside their homes and some were found dead only after several days (IPCC 2007a; Jha 2006; Larsen 2006). Most heat-related deaths in Paris occurred among elderly women (Cadot *et al.* 2007). This did not, however, apply for the Parisian immigrant women, of whom many had roots in Africa and Asia and were living in multigenerational families (Cadot *et al.* 2007). This situation further illustrates the importance of family and other social networks (Table 5.3).

In general, social cohesion and solidarity are important factors because they can build the foundation for establishing local networks and structures (such as local committees for risk reduction) and community-based actions (such as 'community cleaning days' to reduce the risk of waste or branches and other items clogging water channels; see Tables 5.2 and 5.3). Cleaning of waste or avoiding littering in the first place (whether as individual or communal initiative) are measures that address the link between disasters, the urban fabric's physical features and urban ecology (see Section 3.1). Other socially oriented measures relate to education and include investing in children's education or sending children to study outside the settlement. The latter addresses the link between disasters and the urban fabric's physical and socio-economic features (see Section 3.1).

Furthermore, city dwellers also organize themselves to improve communication with different organizations and give the community a more powerful voice in lobbying for better services (including risk reduction and adaptation). In Manchester, UK a spokesperson was elected for this purpose in the areas affected by the 2006 floods (Wamsler and Lawson 2012). Likewise, in informal settlements, such as in Brazil and El Salvador, the interests of residents are often represented by informal residents' associations or local committees.

Another social and/or institutional measure is the fight for legal tenure to obtain access to formal assistance or credits offered by national or local authorities, banks or aid organizations. Legal tenure is important as it means that people are not in constant fear of eviction and are therefore more motivated to improve their risk situation (Wamsler *et al.* 2012) (Table 5.3 and Box 5.1). This is an example of a measure that jointly addresses the links between disasters, the urban fabric, related socio-economic aspects and urban governance (see Section 3.1).

BOX 5.1

The importance of secure tenure for risk reduction and adaptation

Francisca lives in an informal settlement in San Salvador. She does not own the land she is living on and says: 'I do not have a secure house, nor a secure entrance to my house, but if I spend a lot of money on [improving] this, perhaps the next day they come and say "leave, go away from here".' Other people living in the same area describe how government officials are completely ignorant about their situation: 'The government has never had the kindness to visit these remote [meaning informal] places. . .' Consequently, governmental assistance is scarce: 'No, they have not given us anything. We only see them passing by. As we are "private" [meaning informal], as they say it . . . They do not care about us, only for the ones that have formally accessed their land.' (Wamsler *et al.* 2012.)

The importance of secure tenure can also be illustrated with the case of Roberta, a slum dweller in Rio de Janeiro. When asked about how she copes with existing risks, she mentions being a land- and homeowner (as opposed to renting) as a strategy, as well as investing in improvements to her house and land to become less vulnerable to disaster impacts. As she earns her living informally through a local catering business, she does not want to move elsewhere. (Wamsler *et al.* 2012.)

But it is not only in informal, low-income settlements that land tenure can become an issue for people's local efforts and their access to institutional assistance for risk reduction and adaptation. As an example, in the city of Lomma, Sweden, some houses located next to the sea were 'deleted' from the municipal maps (*detaljplan*) (Lomma Kommun 2012). The reasons given were that they are built in a high-risk area, near Öresund and on low-lying terrain, and that the municipality's obligations have 'expired' in the sense that more than 10 years have passed since the granting of the construction permission (*bygglov*) to the homeowners. Whilst it is generally each individual landowner's responsibility to protect the land from flooding, this situation surely impacts the adaptive capacity of the affected Lomma residents.

Finally, city dwellers also adopt emotionally oriented strategies, such as simply accepting or ignoring their risk, seeking emotional support from their social network (family, relatives, neighbours, spiritual or religious group, etc.) or relying on their faith (Table 5.3). As Alberto, a resident of hurricane-prone Caye Caulker, Belize states:

> Just the thought [of my home and possessions being damaged] scares me
> . . . So I hope that doesn't happen. I pray.
>
> (Esdahl 2011: 34)

Faith and beliefs can have both positive and negative impacts on risk. Religious, cultural or spiritual centres can assist people directly and indirectly: by providing resources (both financial and non-financial), access to international support networks of faith, and local knowledge on vulnerable groups and their needs. Furthermore, religious or spiritual leaders can spread the message of the need for adaptation. Religious, cultural or spiritual beliefs can, however, also foster fatalism and inaction if disasters are understood as a result of fate or as punishment for faithless behaviour, and thus not able to be influenced by any risk reduction or adaptation measures. Such understanding often goes hand in hand with a notion of 'blame', which is frequently placed on those already vulnerable or voiceless, such as women or marginalized groups (de Cordier 2011; Reale 2010). A concrete example is presented in Section 5.5. It concerns an informal settler in Zambia who states that, in order to impede climatic changes, women should 'stop wearing trousers' and 'respect their husbands'. Others place their faith in hierarchical structures and thus depend wholly on their assistance, leading to the same fatalistic behaviour. Such a situation is more common in cities in high-income nations (especially when compared with informal settlements in low- and middle-income nations).

5.3 Coping through preparedness for response

> Now I keep sandbags and a floodgate in my cellar. Hopefully I will be around next time [it happens] so that I can install them in time.
>
> (See Wamsler and Lawson 2012; adapted from real situation)

> Make sure you have your documents all in one place; that you know where your money is, [that] you have access to money. Money is more important than anything – money is more important than a can of beans.
>
> (Esdahl 2011: 19)

> . . . our tables are very high and so also are our wardrobes; they are made in such a way that we can climb and sit on top of them.
>
> (Douglas et al. 2008: 197)

City dwellers take a range of different measures in preparation for potential emergencies that allow them to temporarily adapt their behaviour to changed circumstances. Some preparations are made throughout the year, others are only made shortly before a potential hazard impact (e.g. following a warning). Typical examples include storing food and water bottles, or having equipment such as fire extinguishers, a portable hotplate or cooker (in case the kitchen is under water or damaged), air-conditioning appliances or a fan (Tables 5.4 and 5.5). Further flood-preparedness measures include the storage of plastic sheets, sandbags, electric pumps, floodgates, objects that can help to temporarily raise furniture, and other items that can be used to block wastewater pipes and prevent backflow if water levels rise.

Physical response-preparedness measures include temporary improvements such as the above-mentioned floodgates, and temporary or permanent structures that provide a refuge during emergencies. These can take the form of open spaces, squares and elevated platforms, emergency rooms in a building or dedicated emergency shelters (Tables 5.4 and 5.5). It is mainly in marginal settlements with few governmental services where whole communities are involved in the organization or construction of their emergency shelter. In the low-income settlement of Rocinha in Rio de Janeiro, Brazil, some residents even buy or rent an 'extra house', either within or outside their settlement, where they and their families can take shelter, if necessary (Wamsler *et al.* 2012). In the earthquake-prone Kathmandu Valley of Nepal, local tradition means that open spaces (called *la chhi*) are left in the urban fabric for public use to provide places where people can escape falling rubble. These spaces have proved to be crucial in past earthquakes (SAARC 2008) (Table 5.4).

For many physical preparedness measures there are both simple low-tech arrangements using whatever is available (such as stones, bricks, plastic bags, cloths or burlaps) and more high-tech solutions. Examples of the latter are commercial floodgates,[21] tarpaulins for temporary slope protection, water-absorbing and self-inflating 'sand' bags (that inflate when they make contact with water)[22] and backflow prevention devices (see Table 5.4). In flood-prone areas, the purchase and storage of such backflow prevention devices is increasingly recommended by city authorities in the north, such as in Denmark's capital, Copenhagen (Hasling 2012; Leonardsen 2012). Other examples of low- and high-tech measures include the use of burlaps wrapped around shoes versus commercial spike shoes to cope with icy streets, or the dampening of bedclothes for sleeping versus having air-conditioning to cope with heat. Another example comes from Lagos, Nigeria, where residents of the low-lying coastal slum settlement Iwaya/Makoko use buckets to

Table 5.4 Coping strategies for preparedness for response (hazard-specific). This categorization mainly includes individually taken physical measures, and to a lesser extent interrelated social measures.[a]

Hazard	Illustrative examples of activities/measures
Precipitation, tsunami, flood and sea-level rise	Possession of commercial products for flood or rain protection (e.g. floodgates for the main entrance to the property, self-inflating 'sand' bags, tarpaulins for slope protection or backflow prevention devices)
	Having an available room (e.g. higher room on an upper floor) or an 'extra house' to temporarily move to if need be
	Construction of an emergency room for tsunamis (highest floor)
	Storage of objects that allow a temporary increase in the height of furniture and other storage facilities, e.g. putting bricks under furniture legs if need be
	Storage of plastic sheets to put on the roof, on the inside walls or over the bed if need be
	Taking belongings to a more secure location (within the house, for example, on a bunk bed or on a higher platform outside the house) (also ad hoc)
	Being in possession of an electric water pump
	Storage of objects to block wastewater pipes to avoid flooding, backflow and/or related contamination (also ad hoc)
	Construction of makeshift barriers in front of the door (also ad hoc)
Landslides and erosion	Storage of sand, soil, stones, plastic sheets, etc. to temporarily improve slope protection during heavy rains (also ad hoc)
Temperature, heat- and coldwaves	Being in possession of air-conditioning, electric fans and/or back-up heating sources
	Being able to stay or sleep in cooler or warmer places (e.g. having access to a cellar, an air-conditioned car, or a air-conditioned or heated public place such as a supermarket, metro carriage, library or cinema)
	Being able to work from home or to work less (also ad hoc)
	Storage of bottles of water, drinking more water and taking more showers to cool down (also ad hoc)
	Avoiding alcohol, tea and coffee, which have a dehydrating effect (ad hoc)
	Wearing light, loose-fitting clothing (or warm clothing in case of coldwave) (ad hoc)
	Keeping roof/roof terrace wet by regularly pouring water on it (to reduce indoor heat) (ad hoc)
	Storage of de-icing salt
	Use of burlaps or spike shoes, snow tyres, etc. to cope with icy streets

Table 5.4 (continued)

Hazard	Illustrative examples of activities/measures
Drought and water scarcity	Water storage in cisterns, barrels, etc.
	Use of low-tech methods to filter or purify water (e.g. boiling water or using filters made of clay or cloth) (also *ad hoc*)
	Making longer trips to fetch water (*ad hoc*)
	Buying water in bottles or from private tap owners (*ad hoc*)
	Reduced use of water by, for example, showering, washing dishes, doing laundry and flushing the toilet less frequently (also *ad hoc*)
	So-called 'flying toilets', i.e. people defecate in bags and toss them outside their window when they cannot flush their toilet (also *ad hoc*)
Fire	Installation of a fire extinguisher in the house
	Storage of sand to throw on electrical wires that have caught fire (as water will cause them to short circuit)
Wind- and snowstorm	Construction of an emergency room (below ground or on ground level in the centre of the house)
	Temporarily securing objects to prevent them falling down
Earthquake and volcanic eruptions	Leaving doors open so that escape routes cannot become blocked
	Staying at home (to avoid accidents caused by damaged buildings and infrastructure) (*ad hoc*)
	Leaving open spaces in the (self-built) urban fabric which can provide a refuge from falling rubble
	Observation of animals' behaviour (e.g. cats and dogs, which can 'feel' upcoming earthquakes)
	To protect from volcanic ash: shutting down a building's mechanical systems and air-conditioners, protecting air intakes and closing other openings (such as doors and windows) to reduce indoor damage from volcanic ash
	Cleaning volcanic ash from roof to avoid collapse caused by the weight of the accumulated ash (*ad hoc*)

Note:
a The categorization of measures is not always clear-cut and there are no doubt overlaps (see Tables 5.1–5.6).

scoop floodwater out of their houses, while wealthier members of the same community employ mechanical pumps (Douglas *et al.* 2008).

There is further a range of other, non-hazard-specific actions people take to prepare for disaster response (Table 5.5). Many city dwellers strongly rely on institutional assistance but communitarian measures are also manifold, especially in the southern hemisphere. The creation of local emergency groups, for instance, is a common measure in many southern low-income settlements. Many low-income settlements in Central America established such groups following Hurricane Mitch in 1998, often with some support from national and international NGOs (e.g. Wamsler and Umaña 2003).

Similarly, the creation of social cohesion, solidarity and community networks forms the basis for various response and other risk-reducing activities (see Table 5.3). Regarding response preparedness, social cohesion, solidarity and community networks can facilitate mutual aid both during and in the immediate aftermath of disasters. This can consist of door-to-door advice and evacuation procedures, guarding empty houses, transportation of belongings to higher-level streets, the provision of temporary accommodation by community members living in more secure areas, and sharing food and services (such as toilet facilities) (Table 5.5). Close linkages with institutions can strongly assist in such efforts. Institutional buildings, such as places of worship, can, for instance, function as emergency shelters or food distribution sites (which might be 'open' for all affected citizens or only for 'members').

Another important aspect of preparedness for response is the anticipation and monitoring of hazards. In this context, social cohesion and community networks are crucial for the rapid diffusion of hazard information for early warning, especially in marginal areas. The information sources used for the prediction of extreme weather events include television, radio, the internet, the local priest or church, as well as people's own observations and traditional/indigenous monitoring systems. In Los Manantiales, San Salvador (El Salvador), people monitor flood risk by observing the river's water level, the appearance of the clouds and the noise created by up-hill rain (Wamsler 2007a). Similarly, a woman in the Alajo settlement in Accra, Ghana states:

> As soon as the clouds gather I move with my family to Nima to spend the night there.
>
> (Douglas *et al.* 2008: 197).

In Rocinha, Rio de Janeiro (Brazil), the appearance of clouds and the increase in the number of cockroaches entering houses are used to forecast

Table 5.5 Coping strategies for preparedness for response (non-hazard specific). This categorization mainly includes social, economic and institutional measures. Communitarian and hierarchical measures are prevailing.[a]

Strategy[b]	Illustrative examples of activities/measures
Creating family and community support networks	Creation of community emergency groups
	Engagement in community matters and community-based decision-making for preparedness (see also following aspect)
	Interacting and sharing resources with people from the neighbourhood and other actions that aim to increase community solidarity and coordination. In case of a disaster, this facilitates coordination and team work for:
	• Overseeing empty houses and evacuated people who are asleep
	• The transportation of people's belongings to higher-level streets (if rain/flooding is expected)
	• Moving to refuges (e.g. private houses of other community members) in anticipation of a disaster
	• Door-to-door warning to alert at-risk residents, urging them to evacuate
	• Sharing food and services (e.g. toilet facilities) with others during emergencies
	• Storing water in bottles, cans, etc. for community use
	• Accessing cold (or warm) places in the community
	• Collective making and/or storing of items which may help rescue community members during disasters (e.g. locally manufactured boats in flood-prone areas)
	• Early warning
	See also below and Table 5.3 (under social cohesion)
Creating information structures and early warning	Mutual learning and creation of local structures or mechanisms to access and disseminate risk information:
	• Contacting government organizations (at local, municipal and national level)
	• Searching for relevant information (e.g. weather forecasts on the Internet)
	• Going to church (priests are a source of information), asking neighbours, listening to the radio, watching television

(continued)

Table 5.5 (continued)

Strategy[b]	Illustrative examples of activities/measures
Preparing for eventual evacuations	• Observation and monitoring of issues that may indicate disaster occurrence (e.g. colour of clouds, level of rivers and behaviour of animals such as ants, cats and dogs)
	• Video recording of growing crack in the ground (to warn others and to inform authorities and media)
	• Establishing informal communication structures for early warning between neighbours, community members, etc.
	• Using stories, proverbs, songs, poems or festivals to 'store' and disseminate information
	Arrangement of emergency shelter, food distribution points, etc. Spiritual and religious groups can assist in this context by offering churches, temples or mosques for this purpose
	Keeping an emergency store of food, water, torches, etc.
	Storage of a portable cooker (e.g. to use in shelter or at home if the kitchen is damaged or under water)
	Taking family members to a safer place (also ad hoc)
	Preparing food in advance for the children so that, if need be, they can eat and then quickly be sent to neighbours or family members in more secure areas
	Staying awake in order to hear warnings
	Family members staying at home, trying to prevent damage to home and assets
Physical multi-hazard measures	Construction of emergency shelter or emergency rooms within buildings
	Building of emergency facilities, e.g. for food distribution
	Exchange of rooms/apartments so that less mobile people obtain the more accessible ones
	Construction of safe outdoor places which provide protection from falling rubble, or of elevated platforms in flood areas

Notes:

[a] The categorization of measures is not always clear-cut and there are no doubt overlaps (see Tables 5.1–5.6).

[b] In relation to psychological and/or emotional support mechanisms, as well as the creation of linkages with institutions to access formal assistance, see Tables 5.3 and 5.6.

heavy rains. Communities located in the Limpopo River Basin in Mozambique predict floods though observing the activities of ants (Hamza *et al.* 2012) (see Box 6.2).

Tables 5.4 and 5.5 show the wide range of preparedness measures taken in advance of potential emergencies and also include responses that may be taken *ad hoc*. For example, in excessive heat people take several showers during the day, work less, work from home or drink more water to cool down. Some *ad hoc* measures come with major trade-offs and are taken as a last resort. In case of water scarcity, *ad hoc* measures include the purchase of overpriced bottles of water or a reduction in the amount of water used for drinking, cooking, cleaning and washing which may, in turn, create health and hygiene risks (Table 5.4). Living under conditions that allow people to take adequate *ad hoc* measures (such as being able to work from home or take a longer 'siesta' if needed) can thus be important to assist people in their response efforts.

5.4 Coping through preparedness for recovery

> I have a homeowner's insurance which, in theory, covers potential flood impacts, but lately there are a lot of rumours about weather changes and insurance reliability.
> (Helsingborg resident, adapted from real situation)

> [When the 1988 earthquake struck] the only thing on my mind [at] that time was to rebuild my house and start a normal life again. For that I needed money so I worked for others for the money and managed [to get] some amount from my relatives as a loan from them.
> (SAARC 2008: 43)

City dwellers deploy various measures that help them to recover more quickly from disasters than they would without such measures. At individual and household level, such measures are called 'preparedness for recovery' or alternatively 'self-insurance'. To insure themselves, people create formal and informal security systems that grant them access to financial or non-financial post-disaster assistance if need be. Such post-disaster assistance allows them to overcome disaster impacts such as housing damages, loss of income or injury, and swiftly bounce back to their former, or even better, living conditions (Table 5.6).

Financial assistance for post-disaster reconstruction and rehabilitation can come from various sources, including formal bank loans, informal credits from friends or employers, savings accounts, community-based savings schemes, donations, the selling price of assets or the payouts from insurance policies (Table 5.6). Especially in urban low-income settlements, people may also save their money 'under the mattress', work extra hours or take an additional job to increase their income, and stockpile assets such as construction material that can be sold quickly if necessary. This last is a measure that addresses the link between disasters, the urban fabric and urban economy (see Section 3.1). To alleviate financial distress after Hurricane Mitch in 1998, a San Salvador resident sold seven roofing sheets of corrugated iron and then re-roofed his home with an old car body. He had deliberately not nailed down the corrugated iron so that he could resell it, if need be, at a higher price:

> I never nail down the iron sheets that I use for roofing my house so that I can sell them if I want; if I need some money, like [I did] after Hurricane Mitch.

(See Wamsler 2007a)

In general, city dwellers who have access to effective social security systems, offered by public authorities or the private sector, are likely to be better prepared to recover after disasters than those who solely depend on informal structures and services. Being fully reliant on hierarchical structures for social protection also, however, has shortcomings, especially in a context of climate change.

Many city dwellers from high-income countries see insurance policies (e.g. homeowner's insurance) as the most effective risk reduction measure available to them (Wamsler and Lawson 2011).[23] Accordingly, after disaster impacts they look primarily to the insurance industry for compensation. For example, the first reaction of residents in Manchester, UK, who were flooded in August 2004 and July 2006, was to contact their property insurer (Wamsler and Lawson 2011). After the two flood events, insurance claims resulted in payouts of typically between £30,000–45,000. These payments did not, however, cover the cost of temporary accommodation, which, for most people, was required on the two occasions for between three and eight months at a time. In addition, after payout, the property owners faced varying increases in insurance premiums and in the 'excess', i.e. the initial amount of the uninsured loss that they would have to pay. The excess increases varied from zero to £15,000. Premiums went up in some cases, ranging from a modest to a three-fold increase. In some cases further insurance was refused. As the insurer's liability is to replace loss and not to provide improvement, there was little incentive to repair flood-damaged properties with resilient materials because insurers would not pay for that.

Simply moving a fuse box to a higher place on the wall was considered to be an improvement and therefore not covered. In general, very few insurance companies offer payments up to the like-for-like amount, with the policy-holder paying the extra cost of a resilient repair.

The shortcomings of being fully reliant on hierarchical structures for social protection can also be illustrated by the words of a resident of Caye Caulker, Belize, after Hurricane Keith in 2000:

> A lot of people are insured, and they got damaged, and it didn't pay them, so it's just a rip off. So most of the people now, they just quit that insurance.
>
> (Esdahl 2011: 35)

Trying to obtain legal home ownership is another important strategy for 'self-insurance' as it serves multiple purposes in risk reduction and adaptation planning (see Wamsler *et al.* 2012; Box 5.1). In both the northern and the southern hemisphere, many city dwellers invest their life savings (and ambitions) in their homes. A legal home is thus often seen to be the household's main asset. It provides access to formal reconstruction assistance and bank credits, can be sold if money is short, and provides other benefits such as income generation (see Section 3.1). However, in risk areas houses can quickly become blighted and unsellable, and for house owners selling their property without disclosing its actual flood risk can become the last resort. In addition, people with limited financial resources tend not to invest in their housing as this means that any losses can be replaced more cheaply and easily.

Seeking a formal employment is another strategy urban dwellers employ to reduce their vulnerability and be better prepared for recovery. A formal job typically means a more secure income (even when disasters result in many lost working days); access to life, health, unemployment or disaster insurance; a retirement pension; direct post-disaster assistance or credits from employers and other workers' benefits such as a yearly thirteenth month bonus and paid sick leave (Box 5.2). The importance of a secure income is illustrated by this quote from a resident of Caye Caulker, Belize, who describes Hurricane Keith in 2000:

> We didn't work for like a whole month . . . like two months. For two months everyone was fixing up, getting back together, trying to get back.
>
> (Esdahl 2011: 23)

Legal tenure and a formal job can thus enable access to financial post-disaster assistance to overcome disaster impacts and speed up people's recovery. There are even cases where residents of low-income areas in San Salvador have made deals with entrepreneurs to illegally obtain employment certificates, as this enables them to access formal insurance schemes and post-disaster credit even though they are not formally employed (Wamsler 2007a) (Box 5.2).

In addition to the described measures that are aimed at obtaining financial resources if needed, city dwellers also require non-financial assistance to

BOX 5.2

The importance of having a formal job for risk reduction and adaptation

A recent study carried out in San Salvador, El Salvador and Rio de Janeiro, Brazil shows the importance of having a formal employment for coping with disasters (Wamsler *et al.* 2012). Interviewees from both study areas stated that a formal employment allows them easier or cheaper access to:

- Secure income (e.g. job less vulnerable to climatic extremes and variability)
- Post-disaster credits (directly from employers, from banks, etc.)
- Life insurance (for family dependents left behind)
- Pension after retirement or in case of inability to work
- Health insurance (allowing better and cheaper treatment)
- Possibility to take (paid) sick leave (e.g. after disasters)
- Other workers' benefits (such as a thirteenth month bonus, regulated hourly rates, staff security regulations and safe equipment)
- Direct post-disaster assistance from employers (such as construction materials)
- An official address (of the employer) required, for instance, to register children at school.

The importance of these issues can be illustrated with the case of an informal worker living in Divina Providencia in San Salvador who pays into the social security system. He has made deals with entrepreneurs to obtain a formal employment certificate to (illegally) access formal social security mechanisms.

In contrast to those who have formal employment, people working in the informal sector generally do not have access to formal social security mechanisms. In addition, they often take on several jobs to make a living and thus have little free time to engage in (community-based) risk reduction (see Box 5.4).

aid speedy recovery from a disaster. This includes help in the form of custody and fostering of children, and the provision of labour for repairing damages, reconstructing houses or clearing up disaster impacts (Table 5.6). The efficient and safe removal of debris, mud and other disaster impacts is crucial for recovering quickly. Without adequate information, the clean-up can become not only a slow but also a dangerous, endeavour (Box 5.3).

BOX 5.3

Preparedness: knowing how to safely recover

Many urban settlements in Brazil are regularly affected by small-scale floods. In some places, a few hours of heavy rain is all it takes for water to enter the houses. Afterwards, people spend a considerable amount of time cleaning up and are also easily affected by secondary hazards. People who walk in floodwater may step on sharp items or a dangerous animal and may catch a disease. Leptospirosis is, for example, spread by rat urine, which is often found in floodwater. Even after the water has receded, there may still be poisonous animals such as snakes, spiders and scorpions in people's houses as they seek refuge in dry places during floods. Furthermore, people fall sick when eating food that has been in contact with contaminated water or could not be kept cold due to electricity failure.

Knowledge on simple precautionary measures can easily reduce the described risks. Examples of recommendations given by the Brazilian Ministry of Health (SVS 2011) include:

- When cleaning up after a flood, use sturdy shoes, gloves and a mask, or at least plastic bags as gloves and a clean cloth to protect mouth and nose
- Be aware of poisonous animals that may hide in houses, especially in dark places
- To reduce the risk of bites, beat or shake carpets, mattresses, sheets, clothes and shoes carefully outside the house
- Do not let children play or swim in floodwater and avoid walking barefoot
- Always keep the area around the house free from litter
- Keep food safe from contamination and rats by storing it in closed glass or tin containers
- In some cases, drinking water and food can be sanitized using sodium hypochlorite. This also applies to water containers and kitchen utensils that have come in contact with floodwater, the kitchen surfaces, walls and floors.

Staying informed about the risks and secondary hazards of recovery is a vital, but rare, way of being prepared (Box 5.3).

For accessing both financial and non-financial post-disaster assistance, the creation of social cohesion, family and community support networks and good relations with organizations from which one can seek assistance (e.g. governmental agencies, local committees and religious, professional and political groups) is crucial (Table 5.6). This is especially true for marginal and low-income communities. The importance of family support networks is illustrated by this statement from an urban dweller in Maputo, Mozambique:

> During the 2000 floods, I lost everything. My house was destroyed, including the latrine, and everything. That is why I do not have a bed. My neighbours suffered too, but they managed to save their goods. Because of my age and being without a husband, I couldn't remove my goods and leave the area ... I survive because of family support.
>
> (Douglas *et al.* 2008: 198)

Non-financial support mechanisms for coping with post-disaster challenges further include psychological or emotional-oriented strategies, such as seeking assistance to treat post-disaster trauma, resorting to substance abuse as a way to handle stress or trauma from disasters, seeking comfort with family members, friends, or spiritual or religious groups, or relying on religious faith. In the recovery context, local faith communities can assist through their experience of dealing with people's personal crises and grief and access to international networks of faith (Reale 2010) (see Section 5.2 and Table 5.6).

Finally, another preparedness measure is the use of construction materials that easily recover (or can be recovered) after hazard impacts (Table 5.6). For example, wooden plank flooring resists contact with water and is less prone to water-clogging once the water recedes following heavy rainfall (Jabeen *et al.* 2010).

5.5 Differences in city dwellers' approaches: from single coping strategies to the effectiveness of coping systems

The foregoing Sections 5.1–5.4 provide an overview of the measures urban dwellers living in areas exposed to hazards and climate change take to reduce or adapt to associated disasters and risk. On this basis, ask yourself: what are the similarities and differences that exist between the measures taken by urban dwellers living in low-, middle- and high-income nations?

Table 5.6 Coping strategies for preparedness for recovery. This categorization mainly includes individually taken economic measures with strong dependencies on hierarchical or community-bases structures.[a,b]

Strategy	Illustrative examples of activities/measures
Economic diversification	Economic diversification allowing people to increase income during the recovery phase by:
	• Taking on an extra job or a more profitable job (also *ad hoc*)
	• Changing job sectors from one where demand decreases after a disaster (e.g. vending clothes) to one where demand rises after a disaster (e.g. the construction sector) (also *ad hoc*)
	• Working longer hours to increase income (also *ad hoc*)
Assets and investments	Use of construction materials that can easily be recovered after hazard impacts (e.g. wooden plank flooring in flood areas – note that this measure is hazard-specific)
	Saving money or acquiring physical assets that can be easily sold if need be, for instance:
	• Using reusable construction materials, which can be sold and replaced with other objects
	• Stocking up saleable household assets and construction material
	• Owning land or a property/house
	Reduction of household expenses to increase disposable income, e.g. cutting down firewood instead of using gas ovens (also *ad hoc*)
	Deciding not to invest too heavily in housing or infrastructure as this means that any losses can be replaced more cheaply and easily
	Acquisition or maintenance of assets that can serve as collateral for obtaining formal credits (e.g. a formal job, legal tenure, own property/house)
Insurance	Paying into formal insurance schemes
	Participating in informal insurance or savings schemes (e.g. community emergency funds or local savings groups)
	Taking on a formal job or obtaining a certification of formal employment to gain access to related insurance mechanisms

(continued)

Table 5.6 (continued)

Strategy	Illustrative examples of activities/measures
Creating cohesion, solidarity and/or reciprocal relationships	Creation of social networks (with relatives/friends) from whom one can seek recovery assistance in form of: • Borrowing money • Taking bank credits (through family members) • Receiving remittances • Obtaining food, construction materials or labour work (for childcare, reconstruction work, etc.) • Temporarily moving in with other family/community members • Post-disaster clear-up (debris, mud, washing clothes, etc.) See also Table 5.3
Creating linkages with institutions to access formal assistance	Participation in the local executive committee Becoming a member of a political party, professional society or religious and spiritual groups from which one, in case of emergency, can seek assistance Maintaining good contact with NGOs, the local government (municipality) and national government organizations, from which one, in case of emergency, can seek assistance Improving the possibility of accessing post-disaster assistance, in some cases even through the intentional increase of risk (e.g. by renting in a high-risk area or moving into a disaster-affected area where more help is available) (also *ad hoc*)
Precautionary measures during recovery	Knowing what precautions to take during post-disaster rehabilitation and cleaning up. Examples in the case of floods are using gloves and sturdy shoes, taking care not to step on sharp items or poisonous animals that may be underwater, not letting children play in floodwater, and keeping the surroundings free from litter (also pre-disaster)
Psychological and/or emotional support mechanisms	Relying on social networks (family, relatives, neighbours, religious group, etc.) or religious faith for emotional comfort and support during disaster recovery (also *ad hoc*) Seeking assistance to cope with post-disaster trauma (e.g. from mental health professionals or spiritual and religious groups that have experience in dealing with personal crises and grief) (*ad hoc*) Resorting to substance abuse to cope with psychological pressure after disasters (*ad hoc*) See also Table 5.3 under psychological and/or emotional support mechanisms

Notes:

[a] Since there are not many non-hazard-specific measures, both hazard-specific and non-hazard-specific measures are included in this table.

[b] The categorization of measures is not always clear-cut and there are no doubt overlaps (see Tables 5.1–5.6).

There are many similarities in the ways urban dwellers around the world reduce and adapt to different hazards and risk. As shown in Table 5.7, several of the hazard-specific measures are broadly speaking the same, and they are often based on individual or hierarchical (as opposed to communitarian or fatalistic) patterns of social behaviour. Many at-risk citizens, for instance, take individual actions to build retaining walls and small embankments to prevent erosion, landslides or floods; increase doorsill levels in flood areas; improve indoor ventilation or use electric fans to reduce heat stress; temporarily secure objects inside and outside their houses to avoid damages and injuries during windstorms and earthquakes; or obtain formal or informal insurance policies which can compensate for potential disaster losses. In addition, legal house ownership is often perceived as one of the key assets and an important 'buffer' against unforeseen events. This is despite the fact that properties located in risk areas can quickly become blighted and unsellable.

Another similarity that exists in the measures taken in low-, middle- and high-income communities is the fact that citizens seldom feel well informed or supported in their local efforts (Table 5.7). This was also the conclusion of a project on climate change impacts in the Baltic Sea Region called BaltCICA. During a citizens' summit, carried out during the course of this project, citizens explicitly demanded that the city authority provide better advice on what they can (and should) do to reduce their own risk (BaltCICA 2012).[24]

Other similarities in city dwellers' risk reduction and adaptation measures can be identified when contrasting them with rural dwellers' coping strategies. People living in urban areas have often become detached from indigenous knowledge and traditional coping mechanisms to reduce and adapt to disasters and increasing risk (ABC 2012; Audefroy 2011; Khan 2008; Shaw *et al.* 2008). In addition, city dwellers' coping strategies have a comparatively strong focus on housing and land issues and less emphasis on productive sources of livelihood. Their measures are generally also more individually oriented and less deliberate than in rural areas (see Wamsler 2007b). Reasons for this relate to the fact that communitarian strategies, based on solidarity and reciprocity, work best (a) in settlements where people have family members living close by, (b) where there is not too great a disparity in residents' income levels, (c) where family members or other dependents are not simultaneously affected by disaster impacts, and (d) where disasters happen repeatedly, but not too frequently, and have mostly short-term impacts (Morduch 1999). In urban areas, such conditions seldom apply, and even less in a context of climate change (for example, due to increased frequency and intensity of disasters and rising numbers of related rural–urban and inner-city migrations; see below under 'Effectiveness of coping strategies').

Apart from the similarities described previously, there are also many differences in people's risk reduction and adaptation measures when contrasting the situation in northern and southern cities, and especially when comparing formal urban communities of high-income nations with informal urban communities in low- and middle-income nations (Table 5.7).[25]

In the north, the prevalent coping strategy is most likely the procurement of an insurance policy. Northern citizens often see insurance as all the protection they need and the most effective risk-reducing measure available to them (Wamsler and Lawson 2011, 2012). The confidence in formal insurance mechanisms and the relatively low level of further engagement for risk reduction and adaptation is related to people's strong confidence in existing social security, welfare and urban governance systems. In the words of Lykke Leonardsen, Head of the Department of Strategic Planning of Copenhagen City, Denmark:

> Because we have such a good public service, people always expect the public authorities to solve all the problems they are facing [due to increasing disasters and climate change], rather than them taking [risk reduction and adaptation] actions themselves.
>
> (Leonardsen 2012)

As a result, even though about half of Copenhagen residents were affected by a cloudburst in 2011 and experienced damage to their basements, only 15 per cent had taken measures to protect themselves before the next cloudburst season (Leonardsen 2012). In the case of the 2006 floods in Manchester, even after the event only 27 per cent of the victims engaged in some form of measure to reduce their risk (Wamsler and Lawson 2012). Similarly, a study carried out in Helsingborg, Sweden, has shown that the winter storm in 2011 that affected 60,000 households did not lead to many improvements in adaptive behaviour post-event (Lindblad 2012).

In southern cities, social security, welfare and urban governance systems are often less developed and generally exclude the urban poor who live in marginal areas (see Box 5.1). Consequently, the level of people's individual engagement for risk reduction and adaptation tends to be higher. The relatively low level of institutional capacity for risk reduction and adaptation in many southern cities can therefore, to a certain extent, be compensated for by the rich range of innovative local-level measures employed by at-risk citizens. This is especially true for cities that have a long history of dealing

Table 5.7 Similarities and differences between city dwellers' measures for risk reduction and adaptation in low-, middle- and high-income nations

Context	Measures
Similarities: not context specific	Many hazard-specific measures are broadly speaking the same
	Focus is on individualistic and hierarchical (as opposed to communitarian and fatalistic) behaviour
	People feel insufficiently supported in their own efforts by city authorities and other urban organizations
	People feel poorly informed about the means and responsibilities they have to protect themselves more effectively
	Many city dwellers have become detached from traditional coping mechanisms
	In contrast to rural areas, there is stronger focus on housing and land issues and less emphasis on productive sources of livelihood
	Owning a (legal) property/home is often perceived as a key asset and 'buffer' against unforeseen events (a measure which is, however, less and less effective in a context of climate change)
	People are increasingly excluded from accessing (affordable and effective) insurance policies
	City dwellers are increasingly driven to find their own (in parts illicit) solutions (due to inadequate institutional assistance and climate change)
Differences: specific to high-income countries	More confidence in hierarchical structures for risk reduction and adaptation
	Insurance coverage is a predominant measure, often seen as the only effective risk reduction strategy at hand (see also above under 'similarities')
	More high-tech materials and techniques used
	Wider choice/offer of commercial products to reduce city dwellers' risk (e.g. commercial floodgates)
	Trend towards ecological behaviour, which also supports, in parts, adaptive behaviour
Differences: specific to low- and middle-income countries	People often reduce and adapt to risk more deliberately and more actively, especially in informal areas
	Formal insurance does not play a major role, especially for the urban poor
	More low-tech materials and techniques used
	Less use of commercial items to reduce or adapt to risk
	Economic diversification is a more common coping strategy
	People are more frequently engaged in measures at community or settlement level, especially in informal areas
	More community-based actions for risk reduction and adaptation
	More involvement of civil society organizations (NGOs, CBOs) and bottom-up movements
	Labour, time and expenses (relative to income) invested risk reduction and adaptation are more considerable for the urban poor
	Greater variety of approaches taken regarding: (a) measures addressing all types of risk factors, (b) measures addressing both physical and non-physical issues and (c) measures based on individualistic, communitarian and hierarchical behaviour

with disasters and coping with the limitations of national and local government. This situation leads to further differences in city dwellers' ways of coping in relation to the following aspects:

- Materials and techniques for structural improvements
- Use of economic diversification for risk reduction and adaptation
- People's engagement in measures at community or settlement level
- Required workload and relative expenditures for risk reduction and adaptation
- Support from private sector and civil society organizations
- Bottom-up movements
- Effectiveness of coping strategies (see Table 5.7).

Materials and techniques for structural improvements. Better-off citizens generally have better access to highly skilled manpower and specialized products to reduce their risk, such as commercial floodgates, self-inflating 'sand' bags and insulation products (see Sections 5.2–5.3). The market for such products is mainly in the northern hemisphere. In contrast, people with limited resources have to use a greater variety of (often low-quality) materials and techniques, using whatever is available to them. For instance, they insulate their houses' ceilings with cardboard or old cloth, use corrugated iron or mattress springs to hold back soil and mitigate landslides, and build embankments from bricks, stones, filled nylon bags or old car tyres (see Figure 4.3).

Economic diversification. In the southern hemisphere, constructive improvements of people's housing and their surroundings are often complemented with a range of non-physical measures, which address all types of risk factors. Economic diversification for reducing and adapting to different kinds of risk is a common strategy for many southern citizens where social security and welfare governance systems are weak or non-existent (see Table 5.7).

Citizens' engagement at community or settlement level. City dwellers who have to deal with a lack of adequate services and governance structures often see themselves forced to engage in community-based efforts to reduce risk at the community or settlement level. Examples are repairs of public infrastructure such as wastewater pipes to avoid flooding and related contamination, the clearing of objects blocking the flow of rivers, or the removal of waste from slopes to reduce flooding caused by blocked water gutters and open water channels.

Workload and expenditures. A study carried out in San Salvador showed that the urban poor can spend up to 75 per cent (and on average more than

9 per cent) of their income on reducing disaster risk (Wamsler 2007a, 2007b). This is in addition to construction materials that are obtained for free, family members' free labour, the opportunity costs of the considerable time spent on reducing risk and the negative impacts of some coping strategies, such as the high interest paid to informal moneylenders. Post-disaster expenses add up as well: replacement of belongings washed away during floods and landslides, temporary income losses, recovery efforts, and the gradual loss of investments put into the incremental building of housing and community infrastructure (see former aspect). Even when NGOs or other urban actors provide support for assisting the urban poor in improving their settlements, they often require the residents' manpower and thus present a considerable workload for them (Box 5.4 and Figure 5.2).

Institutional support. A variety of international, national and local NGOs, as well as CBOs, support local risk reduction and adaptation in low- and middle-income nations to compensate for their relatively low level of institutional capacity. They either work through national and municipal governments or engage directly with at-risk citizens. Although less frequent, in the northern hemisphere civil society organizations can also be found that aim at assisting people in their risk reduction and adaptation efforts, such as people's response preparedness. An example is the Swedish NGO *Civilförsvarsförbundet* (Civil Defence Alliance), which provides courses to teach households how to cope with different types of emergency situations,

---

BOX 5.4

Institutional support for risk reduction and adaptation demands increased workload for the urban poor

Maria, a female resident living in an informal settlement in San Salvador, with six years of education, takes an active part in the community-based work offered by the Salvadorian Foundation of Development and Housing (FUNDASAL) which is aimed at reducing flood and landslide risk. Although several residents express their reluctance to actively participate, she says: 'It is true that we [meaning the poor] have to work [in order to adapt and reduce increasing risk], but this is how it is, we have to work hard if we really want to make a change here and have a better life.' (See Wamsler *et al.* 2012.)

such as longer electricity blackouts or contaminated drinking water, and how to make their own risk analyses.[26]

Bottom-up movements. With more community engagement and the increased involvement of local NGOs and CBOs in the southern context, there are comparatively strong bottom-up movements for risk reduction and adaptation (e.g. shack/slum dwellers' federations; see Section 5.6). In the north, such interest and bottom-up movement for risk reduction and climate change adaptation does not exist. It does however exist for climate change

Figure 5.2 Urban poor in San Salvador, El Salvador engage in slope protection with support from a local NGO. Source: Christine Wamsler.

mitigation. In fact, in several northern (and especially European) countries, citizens are increasingly changing their habits towards more environmental or ecological behaviour (European Commission 2008). Related action taking supports local climate change mitigation and pushes governments to act on reducing greenhouse gas emissions. The importance people give to the role of the environment 'translates' into actions that may not always bring direct advantages but are based on the consideration of the well-being of 'future generations'. According to the European Commission (2008), citizens of the European Union commonly contribute to environmental protection with small day-to-day actions such as recycling waste, reducing waste (for example by buying bigger sizes, concentrated products and second-hand items or by avoiding the purchase of over-packaged products), cutting home-based energy and water consumption, using their car less, and buying products that are locally produced and environmentally friendly. So-called 'climate change champions' (i.e. citizens committed to climate change mitigation) take an active role in spreading the word on how people can employ such day-to-day actions to reduce their carbon footprints, and also in trying to influence governments to become more active in climate change mitigation.[27] Although in the north similar engagement and bottom-up movements do not exist for adaptation, some aspects of environmental or ecological behaviour can also foster urban resilience. For example, a more sparing and efficient use of water increases the resilience to drought and water scarcity. This is also true for other resources, like energy, whose availability may be reduced or uncertain under future climate conditions. In addition, reducing and recycling waste has the potential to reduce flood risk and related contamination.

Effectiveness of coping strategies. There are strong differences as regards the effectiveness[28] of city dwellers' coping strategies, including forward-looking solutions as well as short-lived and even harmful measures. Coping strategies that work well at an individual or household level may be counter-productive when considering the 'bigger picture'. As an example, during drought and hot weather some people cope with water scarcity by storing large amounts of water taken from public sources (see Table 5.4), increasing their number of water wells (see Table 5.2) or drenching their rooftops with water to reduce the indoor temperature (see Table 5.4). Such measures inevitably result in increased water consumption, growing pressure on water supplies and sinking water tables. Similarly, the increased use of fans and air-conditioning during hot weather (see Table 5.4) augments energy consumption and thus contributes to climate change. Furthermore, in marginal settlements it frequently causes informal electrical connections to short-circuit, resulting in power outages or fire which can spread quickly in densely built settlements. Another example is so-called 'flying toilets' which are used in informal settlements throughout the world. When toilet facilities are not available or cannot be flushed due to water scarcity, people relieve themselves into a plastic bag and toss it out the window (see Table 5.4), at great risk to both the environment and public health (UN Water 2007). Notably, the use of 'flying toilets' is also linked to gender-based violence – it is a common strategy for women to avoid the risk of sexual violence when seeking a public toilet or another place outside one's home to relieve oneself (Amnesty International 2010). Other examples include flood defence walls and landfill (see Table 5.1), which can increase downstream flood risk. Landfill can also destabilize properties and pollute the environment, as is illustrated by the following statement from a resident of Caye Caulker, Belize:

> In years gone by, people used to use household waste to fill their property. . . . 20–30–40 years ago we didn't really realize how much that damages the environment. But [some] people continue to do it.
>
> (Esdahl 2011: 45)

Other coping strategies seem to be ineffective in both the short and long terms and at all levels. These include passive behaviour (i.e. a total reliance on hierarchical structures or a belief in divine forces) (see Table 5.3), resorting to drugs and alcohol for psychological coping (see Table 5.6), and dysfunctional measures such as using corrugated iron as retaining walls or roofing houses with loose corrugated iron weighted with heavy objects which can endanger neighbours during windstorms (see Table 5.2). Other examples are traditional beliefs or social relations that lead to the exclusion of certain groups (SAARC 2008), as illustrated by this quote from an informal settler in Zambia:

> The frequent heavy rainfalls that come year after year, and the heat, including the sudden shifts between the two weather conditions are clear signs of a curse. Women must stop wearing trousers, playing football, boxing and going to taverns. They should respect their husbands.
>
> (Simatele 2010: 15)

The exclusion of certain groups from decision-making related to risk reduction and adaptation generally creates barriers to sustainable risk reduction, adaptation and transformation. Traditional social relations and hierarchies that subordinate women are particularly worrying as in many cultures women have exclusive knowledge that is unknown by others in the community (Shaw et al. 2008).

Coping strategies that may be effective in the short term but are likely to prove ineffective in the long term include borrowing from moneylenders at high interest rates (see Table 5.6), selling off assets cheaply in the post-disaster period (see Table 5.6), spending money on temporary arrangements (e.g. short-lived water barriers and channels), cutting down trees for firewood to save money (see Table 5.6), or covering slopes with plastic sheets that pollute the environment and blow into rivers and block them (see Table 5.2).

Coping strategies that, in combination with other measures, may be effective in both the short and, more importantly, the long term include cooperating with neighbours and local committees (e.g. for mutual help or early warning) (see Tables 5.3, 5.5 and 5.6); learning from friends and others (see Table 5.3); growing suitable plants and fighting deforestation in order to reduce heat, flood and landslide risk (see Tables 5.1–5.2); accumulating assets for use as collateral or for sale in post-disaster times without making a loss (see Table 5.6); reducing unnecessary expenses (see Table 5.6); accessing safe and convenient saving arrangements, loans with favourable conditions or insurance mechanisms to which access is conditional on risk reduction (see Table 5.6); implementing physical measures that are not rigid but incremental or flexible (such as detachable roofs as fire breaks [see Table 5.2] or floating houses which rise with the water level [see Table 5.2]); improving waste and wastewater management (see Table 5.2); increasing involvement in decision-making for adaptation planning (see Table 5.3); and investing in or increasing access to formal education (see Table 5.3).

Finally, there are strategies that are mainly concerned with the 'bigger picture' and not with direct gains, such as switching to a (partly) vegetarian diet because meat production requires large amounts of water and energy in comparison with grains and vegetables (UN Water 2007). Dodman and Mitlin (2011) also argue that there is clearly a current interest in addressing the issue of climate change even in low-income communities. The argument is supported with examples from settlements in the Philippines, Tanzania and Zimbabwe where, for instance, communities engage in planting fruit trees to reduce deforestation or limit the settlements' access by car to reduce contamination and provide areas for vegetable cultivation and children's play.

Despite the wide range of coping strategies used, they are often insufficient to keep pace with increasing disaster impacts. There are several reasons for this. At times, institutional assistance creates barriers for effective coping. Examples are programmes that impose risk reduction and adaptation measures which cannot be locally maintained (such as retaining walls built with modern techniques that replace local ones); the development of (post-disaster resettlement) programmes that deprive people of their livelihoods without providing new alternatives; the construction of physical risk reduction measures which provide people with a false sense of security; or the provision of social housing which the homeowners cannot use as a guarantee when applying for credits to improve structural disaster resilience (Wamsler 2007a).

The effectiveness of coping strategies and the success of bottom-up movements relates furthermore to the potential and limitations of community-based risk reduction and adaptation in general. Limitations relate to the fact that (a) they are local in focus, while risk is shaped by factors operating at larger scales, (b) coping is based on past and therefore not on unforeseen events, (c) coping often ignores structural issues leading to political and institutional shortcomings, and (d) urban communities are very heterogeneous.

In addition, the rate and magnitude of current environmental change has various negative impacts on the effectiveness of coping strategies. Before continuing to read, ask yourself: how can climate change impact the effectiveness of urban dwellers' coping strategies?

Effectiveness of coping strategies – in a context of climate change. In a context of climate change, people are faced with a new starting point from which they must cope. Heavy rains are, for instance, increasingly frequent outside the rainy season, which makes floods and landslides harder to predict and reduces the time available for preparations and recovery. There is also a greater need to cope with more frequent and severe hazards; this consumes resources that would otherwise be available to meet subsequent coping needs. It also brings challenges for better-off community members, which can ultimately lead them to opt out of solidarity- and community-based mechanisms. In the words of a resident of Caye Caulker, Belize, who is the owner of a hotel where at-risk neighbours frequently seek refuge during hurricanes:

> The people affect us more, because they are not prepared, they can do more damage than the hurricane.
>
> (Esdahl 2011: 15)

The statement that people can do more harm than the hurricane itself is based on the fact that disaster victims often overstay their welcome, which for the hotel owner means an increased workload and stolen goods, but also a feeling of responsibility that prevents the hotel owner from leaving the island during emergencies (Esdahl 2011).

Furthermore, climate change can make traditional knowledge and conventional coping strategies either obsolete or no longer viable (Shaw *et al.* 2008). A telling example comes from a Mozambican community that had always been able to predict floods based on the behaviour of ants. In 2000, unusual cyclone activity resulted in floods which came so rapidly that there was no time for the ants to react, resulting in more than 700 fatalities (see Box 6.2). Furthermore, quickly changing conditions outpace the ability of locals to adapt through a process of testing and modification. To make matters worse, climate change is creating local conditions that make community-based actions less and less viable. The trend towards individually oriented measures is thus likely to become stronger in urban communities, which reduces adaptive capacities at household and community level (see beginning of this section).

Not only traditional, but also 'modern' coping strategies are increasingly challenged by climate change. In both the northern and southern hemisphere, those at risk are increasingly excluded from accessing (affordable and/or effective) insurance policies, are less and less able to use their home as an asset to cope with shocks and stresses, and even feel in some cases driven to find illicit solutions to their situation. This includes making unauthorized connections from drains to sewer systems, selling houses without disclosing the actual risk and gaining illegal access to insurance (see Box 5.2). Hence, increasing climate change and disasters are bringing the situation of city dwellers worldwide into greater convergence.

If sustainable transformation for risk reduction and adaptation is the aim, it is thus crucial to take urban dwellers' own efforts into consideration and support effective coping. It is, however, not only the effectiveness of individual coping strategies which is determining.

While the effectiveness of individual coping strategies is important, it cannot be taken as an indicator either of individuals' or households' adaptive capacity or of the relevance of particular strategies in enhancing resilience and transformation. What can be a short-term measure for one person or household can be a sustainable solution for another, depending on (a) the context and conditions and (b) the set and combination of strategies. For example, income diversification might mean increased working hours and burnout for one woman who is already overloaded with other tasks. However, for another woman it might translate into increased independence, money, a retirement pension, insurance cover and better social networks. Another example is selling or renting out rooms. This may only help to maintain the 'status quo', or even leave households worse off if they receive less money than they have invested or have to move family members to higher-risk areas. Conversely, for another household, selling or renting out rooms might generate the resources needed for sustainable change, such as the construction of better or safer houses, or investments in health, education and business (see Simatele 2010). Furthermore, while a single coping strategy might not be particularly effective, it can be a vital complement to other strategies, which together create a sustainable coping system.

It is thus city dwellers' combined set of strategies that has to be looked at when comparing and possibly supporting localized approaches. In contrast to the effectiveness of single coping strategies, the effectiveness of coping systems has a broader scope with direct links to the notion of sustainability. A sustainable coping system can be understood as a system that is flexible and inclusive enough to assist individuals, households and urban communities in reducing their risk, while maintaining or enhancing local adaptive capacities both now and in the future so as not to compromise the ability of future generations to meet their own needs.

Both the flexibility and inclusiveness of coping systems are attributes that are crucial in a context of climate change and uncertainty. Flexibility firstly relates to the number of measures that address each risk factor and thus to the redundancy in the coping system (see Sections 2.1 and 2.7). Secondly, it refers to the diverse types of measures that address each risk factor: their thematic and hazard foci and their underlying patterns of social behaviour (see the introductory section of this chapter). Inclusiveness relates to the use of not just some but all of the four potential risk reduction and adaptation measures to ensure that all types of risk factors are addressed (see the introductory section of this chapter and Box 2.4).

An important factor that can positively influence both the flexibility and inclusiveness of people's coping systems is formal education. Increased levels of formal education can provide the skills and social capital that enable people to flexibly find appropriate and forward-looking solutions to climate-induced problems as they arise, and react to anticipated risks that have not yet become manifest (Lutz 2010; Striessnig *et al.* 2013; Wamsler *et al.* 2012). A recent study carried out in the context of the project 'Forecasting Societies' Adaptive Capacities to Climate Change', funded by the European Research Council,[29] concluded that an increased level of education tends to translate not only into more coping strategies, but also into more flexible and forward-looking measures (Wamsler *et al.* 2012). Formal education was found to have a positive influence on the following factors, which are particularly relevant to efficient coping:

* Having a formal employment (see Table 5.6)
* Opportunities or interest in moving to a lower-risk area within or outside the settlement (see Table 5.1)
* Awareness, identification and understanding of current risk levels (see Table 5.5)

- Access to information on risk, risk reduction and adaptation measures (see Tables 5.3 and 5.5 and Section 5.6)
- Use of coping strategies that are directly related to education (see Table 5.3 and Section 5.2)
- Communication of knowledge on risk, risk reduction and adaptation to other community members and the authorities (see Tables 5.5–5.6 and Section 5.6)
- Acceptance and appropriate use of institutional support (see Table 5.3 and Box 5.4). See also Boxes 2.1, 5.2 and 5.5 as regards all aspects listed.

5.6 Citizen-to-citizen (and -city) knowledge transfer

In any strategy that aims to help those most at risk optimize their adaptation to changing hazard and climate change conditions, knowledge transfer is a vital component (IPCC 2000). In this context, most interest is given to the support of horizontal flows between cities (see Boxes 2.12 and 2.13). Less attention is given to citizen-to-citizen flows and vertical connections, such as city-to-citizen or citizen-to-city flows (see Section 4.4). Consequently, urban dwellers' local knowledge of risk reduction and adaptation and its transfer are poorly documented and are generally not taken into consideration in the work of municipal authorities (e.g. Banks *et al.* 2011; Carmin *et al.* 2012; Satterthwaite *et al.* 2007b; Shaw *et al.* 2008; UNISDR 2012b; see Sections 4.1–4.3). Urban institutions rarely know of people's own efforts in the areas where they operate, while urban dwellers often feel poorly supported in their local efforts and uninformed about the means and responsibilities they have to protect themselves (see Table 5.7).

To deal with the lack of institutional support and information, city dwellers develop alternative information channels that enable horizontal knowledge transfer. Citizen-to-citizen knowledge transfer is an inherent part of many coping strategies presented in the Sections 5.1–5.4. The transferred knowledge is the sum of known facts, learned from experience or acquired through observation and study. Examples are local knowledge of storm routes, wind patterns, cloud formations, rain corridors or animal behaviour, which enables people to make preparations and reduce their risk. Taking wind patterns into account in the design of buildings can help to reduce damage. Knowing the colour of clouds that carry hailstones can provide an early warning to run for cover. Similarly, the height of birds' nests near rivers or the presence of certain plant species that indicate a low (or high) water table can assist in selecting adequate construction sites (see Table 5.3) (Singh 2011).

BOX 5.5

City dwellers' coping strategies and education

Increased levels of formal education can provide the skills and social capital that enable people to flexibly find appropriate solutions to climate-induced problems as they arise. Ana, a 40-year-old single mother with 11 years of education, lives in a Brazilian *favela* (slum) and is taking university-entry exams to study journalism. She has not received any institutional support to improve her living situation but she managed to obtain a stipend from the renowned, private language school *Cultura Inglesa* for her son to study English. When asked about how she copes with disasters and associated risk, she mentions a range of different strategies including:

- Regularly looking for risk information on the internet (such as weather forecasts and measures to reduce disaster risk)
- Investing in the structure of her house
- Improving the electricity connections inside the house (distribution and outlets)
- Staying informed about how to prevent diseases like dengue and tuberculosis
- Not throwing trash in the streets (as a strategy for both flood and disease prevention)
- Avoiding hazardous areas during emergencies/evacuations (such as a nearby gas station)
- Sending her son to swimming lessons
- Sending her son to study outside the *favela*.

When asked about her interest in moving to a more secure area, Ana states that there is a difference between *living* in a *favela* and *being* the *favela* (referring to the stigma of its residents), and then highlights that she only lives here because she does not have the opportunity to live anywhere else.

Ana is also working as a health agent in the community to promote preventive health measures, while her ambition is to retrain as a journalist. This supports the understanding that, as educational levels increase, individuals tend to develop a stronger sense of civic duty and a greater interest in politics, which influences their adaptive behaviour and positively influences democracy in general (see Boxes 2.1, 3.4 and 5.4; Lutz 2010; Wamsler *et al.* 2012).

People learn from their neighbours, friends, parents and ancestors. City dwellers' knowledge on risk reduction and adaptation, which is rooted in indigenous knowledge, local culture and traditions, has been handed down from generation to generation (see Australian Red Cross 2010; Shaw *et al.* 2008). For example, in Spain the underground cisterns commonly used for collecting rainwater, known as *aljibes*, were already introduced during the Moorish period in the twelfth century (van Wesemael *et al.* 1998). The white-washed houses that are traditional in many warm countries, such as Greece, are a centuries-old strategy for coping with heat (Ben Cheikh and Bouchair 2008). In Nepal, information about local hazards is also spread by folklore and poetry. The poem *Lokmanjari*, a homage to the devastating earthquake of 1934 which struck Kathmandu Valley, contains detailed descriptions of the causes of vulnerability and advocates local technology for increasing resilience (SAARC 2008). There are many examples of how houses built with traditional materials and technologies, adapted over centuries to the local environment, have better withstood disaster impacts than the often preferred 'modern' houses (Audefroy 2011; Khan 2008; Shaw *et al.* 2008).

The dreadful impact that can result from the lack of transfer of traditional knowledge is illustrated with the following quote from a Thai architect after the 2011 floods in Bangkok in which over 800 people died:

> Bangkok used to be very much a floating city . . . There were raft houses, floating houses, along the river and canals. I think the fact that we have forgotten or we have lost our aquatic instinct is costing us quite dearly.
>
> (ABC 2012)

In order to deal with fast-changing living conditions caused by climate change and urbanization, city dwellers are also developing new information channels that allow horizontal knowledge transfer between the inhabitants from different urban communities, such as internet-based forums or online groups. An example is the *Inteligencias Colectivas* (Spanish for 'collective know-how'). It is an internet-based horizontal learning system and a free database, which collects and promotes technical know-how on informal construction solutions, many of which relate to risk reduction and adaptation.[30] The way people learn and communicate determines the transfer of knowledge on risk reduction and adaptation (DFID n.d.) and certainly evolves over time. Both learning and communicating knowledge on risk and risk reduction and adaptation are positively influenced by people's level of formal education (see Section 5.5 and Boxes 2.1 and 5.5).

In addition to the knowledge transfer between individual city dwellers there are cases where whole settlements have organized themselves to exchange knowledge with communities located in other urban areas. An example is the network of CBOs called Shack Dwellers International, which exists in 33 countries in Africa, Asia and Latin America.[31] This network has created horizontal south–south information flows that increase urban communities' knowledge of issues such as disaster-resistant building practices, waste management and drain maintenance (Dobson 2011). International and national NGOs also support knowledge exchange at community level. An example is the Salvadorian NGO called FUNDASAL[32] which organizes visits and exchanges between at-risk communities in San Salvador.

Examples of city dwellers' local knowledge on risk reduction and adaptation being successfully transferred and integrated into the work of municipal authorities are not very common (Carmin *et al.* 2012; Satterthwaite *et al.* 2007b; UNISDR 2012b). Exceptions are places with strong local networks, associations and bottom-up movements, such as the previously mentioned CBO Shack Dwellers International. Apart from creating horizontal information flows between city dwellers, Shack Dwellers International has also created a platform for information sharing and cooperation with city authorities. It collects invaluable information during citywide enumerations in each of its cities of operation. Once this is processed, it works with city authorities to assure that this data assists in generating targeted and efficient strategies for climate change adaptation that can be jointly implemented with local communities (Dobson 2011).

In general, intermediaries (such as local CBOs, NGOs, networks or associations) are helpful to strengthen the often weak link between community members and technical or political institutions at city level. They can assist in supporting dialogue, mediating the exchange of information, and interpreting technical information so that it becomes intelligible to its 'users' (DFID n.d.). There are, however, also success stories of direct knowledge transfer between citizens and city authorities. An example is the 'Eco-City Augustenborg' project in Malmö, Sweden (described in 'Test yourself' scenario 4.3), where the city authority identified and integrated innovative ideas of local residents in the design of the stormwater system.

5.7 Summary – for action taking

Throughout human history, people have adapted to their changing environment. This accumulated capacity is critical in addressing increasing risk and disasters (including climatic extremes and variability). People's coping strategies are the 'visible' expression of their *used* adaptive capacity. They consist of both planned and *ad hoc* endeavours taken before, during and after hazard impacts (Table 5.8). The current environmental changes faced by humanity are, however, happening at a previously unseen rate and magnitude and are deeply intertwined with complex processes, such as urbanization and globalization. Although these are global processes, they are not uniform. City dwellers' hazard exposure, their vulnerability and related adaptive capacity are thus not equally distributed worldwide and there are huge variations in these three elements within every single settlement, every housing unit and even every household (see Box 5.6).

Whilst city dwellers have substantial adaptive capacities to deal with disasters and changing patterns of risk (see Sections 5.1–5.4), they also have to balance multiple priorities and their coping strategies have limitations, particularly in a context of climate change (see Section 5.5). For example:

> One of the predictions in global warming is that temperatures will rise well above yearly averages to the point where the overuse of air conditioners may shut down the power generation systems in a city, thus leaving neighbourhoods literally to their own devices.
>
> (Pizarro 2009: 38)

Such circumstances occured in 2012 in Washington, DC, the capital city of the United States. Heat and windstorms and the resultant power outages struck together with dreadful outcomes – as described by Bill Frohna, Chairman of the Emergency Department at MedStar Washington Hospital Center:

> Before the storm came we saw some heat-related stuff, but once you throw the power issue on top of the heat, families didn't know what to do . . . When no one has power they don't have a backup plan.
>
> (Dorell 2012)

In a context of climate change, urban dwellers are less and less able to (a) access affordable insurance policies, (b) use their homes as an asset to cope with shocks and stresses, (c) receive adequate institutional assistance in terms of protective infrastructure and information and support for reducing and adapting to risk and disasters, and thus (d) can even feel driven to find illicit solutions to their situation. This last includes making unauthorized connections from drains to sewer systems, selling houses without disclosing the actual flood risk and gaining illegal access to insurance.

Without adequate support from city authorities, aid organizations and other urban actors, urban dwellers fall short of using their capacity to its full extent. In order to support city dwellers' adaptive capacity in risk reduction and adaptation, it first needs to be assessed. Coping and adaptation strategies can be assessed in relation to their objective (i.e. hazard reduction and

---

BOX 5.6

City dwellers' differential patterns of risk[a]

In the context of the project entitled 'Adapting cities to climate-induced risks: a coordinated approach', staff from Gothenburg municipality, Sweden participated in a risk assessment exercise for heatwaves.[b] During this exercise, the participants came up with the following fictive (but possibly in their municipality existent) at-risk citizens.

Dimitris, 67, lives in an apartment on the ground floor in Kvillebäcken (Kville River). He suffers from breathing difficulties and heart failure but gets by without help from home care. He regularly attends the health clinic. His daughter helps him with grocery shopping because he has trouble walking and does not own a car.

Märta, 99, lives in a run-down house in Kärrdalen (Marsh Valley). She has poor eyesight and therefore does not watch the news on TV and cannot read her mail without help. The home care service visits Märta three times a day.

Harald, 44, works as a senior manager at an IT company. He lives in a new condominium in Sannegårdshamnen (Sannegård Harbour). He earns good money but does not have much free time and he leads a stressful and hectic life.

Jesper, 3, lives with his mother and father in a small apartment in Brämaregården (Brämare Yard). Both of his parents receive welfare. Jesper's father has a drinking problem and his mother suffers from mental problems. Neither of them has any contact with their parents or other relatives.

Dilsa, 29, is a single mother of three young children. Dilsa and the children live in a rented apartment in Slätta dam (Slätta Pond). Dilsa works part time as a cleaner and finds it difficult to make ends meet for the entire month. One of Dilsa's children suffers from diabetes and asthma. Dilsa has neither a driver's licence nor a car.

---

[a] Using climate change jargon, one could also say 'city dwellers' differential vulnerability' (see Section 2.1 and Figure 2.4).

[b] The research project was coordinated by the University of Gothenburg. See http://www.gvc.gu.se

Table 5.8 Analysing coping strategies before, during and after hazard/ disaster impacts – including planned and *ad hoc* endeavours. Some examples in case of floods are scribbled in the table to illustrate its use as an analysis framework (see Section 2.1, Tables 2.4–2.5).

Before	Shortly before	During	After (⇒ Before)
Increase height of plinths, doors, sill levels	Increase height of furniture (temporarily, e.g. with stones)	Use services from unaffected neighbours	Help from community members to rebuild
Construct drains/improve drainage system of individual houses	Move family and/or assets to a safer place	Community organizes to clear and improve drainage system of settlement	Borrow money to tackle the hardship
Build higher storage facilities	Store food and water		Move permanently to a new location
Remove/relocate service lines	Make temporary barriers in front of the door		Change building materials

avoidance, vulnerability reduction, response and recovery preparedness, as well as *ad hoc* response and recovery measures; see Tables 2.4–2.5). In addition, they can be analysed with regards to: their thematic foci (i.e. physical, environmental, socio-cultural, economic and political/institutional); their underlying patterns of social behaviour (i.e. individualistic, communitarian, hierarchical and fatalist); their timing (i.e. pre- or post-disaster, taken well in advance or shortly before hazard impacts [Table 5.8]); whether they are planned or *ad hoc*, deliberate or unintentional, carried out with or without institutional support; and the information channels through which they are communicated.

Only an in-depth analysis of people's used and unused capacities can allow urban organizations and planners to encourage or scale up existing coping strategies, and offer new or alternative strategies where needed. Examples and lessons of how this can be done in practice can be found in both the southern and northern context. Naturally, not all commonplace local measures can or should be supported. Careful attention should be given to the effectiveness and sustainability of assistance. Hence coping strategies further need to be assessed by dividing them into:

- Those that can increase adaptive capacity in both the short and long terms
- Those that increase adaptive capacity in the short term but reduce it in the long term
- Those that reduce adaptive capacity in the short term but increase it in the long term
- Those that reduce adaptive capacity in both the short and long terms.

The focus of this division is not on individual, but on community gains. Identifying which coping strategies are effective or do not work well and why is a major challenge. Whilst the use of electric fans might, for instance, be an effective way to cope with heat in the short term, longer-term adaptation consists rather of planning adjustments for improved ventilation and shading, together with the use of alternative energy sources and back-up systems at community level (see Section 5.5). Causal loop diagrams of different levels of detail can help in the process of identifying strategies that are effective and sustainable by providing an understanding of the urban context, inherent urban–rural linkages, people's needs and their local coping efforts (see Chapter 3 and Figure 3.6).

While the effectiveness of individual coping strategies is important, it cannot be taken as an indicator either of the adaptive capacity of individuals or households, or of the relevance of particular strategies in improving resilience and transformation. The success or failure of urban societies in building resilience and moving towards sustainable transformation depends not on the effectiveness of *individual coping strategies*, but on the flexibility and inclusiveness of *coping systems* at the individual, household and community level (i.e. the combined set of strategies). In other words, it is not fruitful to look only at single coping strategies because their short- or long-term effectiveness is dependent on (a) the context and conditions and (b) the set and combination of strategies. A particular coping strategy can be a short-term solution for one household and a long-term solution for another. Furthermore, while a single coping strategy might not be particulary effective, it can be a vital complement to other strategies, which together create a sustainable coping system. Discouraging a supposedly short-term coping strategy without considering the whole (household or community) system might consequently result in reduced adaptive capacity.

In view of this, it is crucial to support the ability of communities to negotiate their needs and rights in order to transform existing coping systems by increasing the flexibility and inclusiveness of these systems, and making them more viable in today's context. Coping should not automatically be seen as maladaptive (or a conflicting practice for sustainable risk reduction and adaptation). It is further not helpful to differentiate between surviving and adapting by only looking at every individual coping strategy in isolation. The analysis of coping strategies also requires a thorough consideration of the combined system set up to cope before, during and after hazard occurrence. On this basis, measures for supporting local risk reduction and adaptation may include improving and accelerating communities' inherent learning mechanisms, increasing people's level of education,[33] and more end-of-pipe solutions such as encouraging or scaling up existing strategies, or offering new or alternative strategies where needed. In line with this, sustainable transformation might be a set of incremental improvements that together, and possibly in combination with new attributes, transform current coping systems from within.

Altogether, if put into practice, the in-depth analysis (of coping strategies, coping systems and unused capacities) enables city authorities, aid organizations and planners to take advantage of existing local capacities and information channels. This, in turn, allows them to: provide assistance that appeals to the various perspectives and efforts of city dwellers; ensure that measures are context-specific, get implemented and are maintained; extract principles that can be transferred to other locations; and ultimately build flexible and inclusive structures for risk reduction and adaptation. In this context, urban actors' legal responsibilities for taking adaptive measures require thorough attention because they can obstruct related action taking (see Sections 4.2 and 6.2). The following issues have to be taken into consideration:

1. Which actors in society have a legal responsibility to take risk reduction measures and respond to climate change?
2. Where are the limits and borders drawn between institutional and individual responsibilities?
3. How can institutional and individual capacities be supported and developed within the existing frames of legislation?

On this basis, the need and opportunities for revising existing frames of legislation can be explored to allow (a shift towards) a more distributed urban risk governance system where people's and institutions' efforts can better support and complement each other. This requires not only incremental, but also transformative improvements to provoke systems change for long-term sustainability.

5.8 Test yourself – or others

The following questions can be used to test yourself or others. The answer to most questions can be found explicitly in Chapter 5.

1. How do you define the term 'coping strategy' and how does your definition relate to the concept of adaptive capacity?
2. How can the measures people take to reduce and adapt to disasters and increasing risk be analysed in a systematic way? Describe in detail how you would structure such an analysis.

3. Provide four examples of measures urban dwellers can take to cope with heatwaves and discuss their effectiveness at individual, household, city and global levels.
4. Describe four measures urban dwellers can take to prepare for speedy recovery from floods, for instance in a marginal settlement in Brazil.
5. Give examples of (a) individualistic, (b) communitarian, (c) hierarchical and (d) fatalistic behaviour of citizens to deal with the risk of earthquakes.
6. What differences are likely to exist between the coping strategies of residents in formal high-income and those in informal low-income settlements?
7. How do effective social security, welfare and urban governance systems relate to people's coping strategies?
8. Provide illustrative examples of risk reduction and adaptation measures that could be taken by urban dwellers living in flood areas. Describe at least one high-tech and one low-tech measure for both vulnerability reduction and preparedness for response.
9. Why is it crucial for city authorities and other urban actors to consider people's local coping strategies in their work on urban risk reduction and adaptation planning?
10. Discuss how the following aspects can have an influence on urban dwellers' local coping capacity:

    a) Having a formal job
    b) Economic diversification
    c) Secure tenure
    d) High level of formal education.

11. Give examples of single coping strategies that may be effective in the short term, but are ineffective in the long term. On this basis, discuss why ineffective coping strategies might be important for individuals', households' or communities' coping systems (i.e. their combined set of used strategies).
12. Name a coping strategy that addresses the link between the urban fabric's physical features and non-physical features of:

    a) the urban ecosystem
    b) the urban society and culture
    c) the urban economy
    d) the urban governance system.

Case studies might also be a good basis for assessing the knowledge presented in this chapter.

Test yourself scenario 5.1: evaluating people's adaptive capacities (northern context). The Municipality of Gothenburg in Sweden has formed a task force for identifying the most vulnerable people within the city. You are a member of this task force. An external consultant has come to facilitate the first meeting. During this meeting, the consultant asks you to come up with a description of five fictive (but possibly real) people living in your municipality, who could be considered at high risk from potential heatwaves. The result is presented in Box 5.6.

Task:

1. Analyse the description of the five residents in a systematic way in order to explain why they can be considered to be particularly at risk.
2. Make up another resident who, in contrast to the described residents, could be considered to be at low risk.
3. Make up another four fictive people, of whom two are living in urban areas and two in rural areas, in order to demonstrate how the characteristic urban/rural features and related urban–rural linkages can influence people's level of risk.

Test yourself scenario 5.2: evaluating people's adaptive capacities (southern context). You are given the assignment to evaluate people's local capacity for risk reduction and adaptation in a small informal settlement in Caracas, Venezuela. The settlement is made up of some 100 houses that have been built directly on the slopes of the El Avila Mountain after most of its natural vegetation had been chopped down. Most houses are made of bricks and cement but some, in the very highest and steepest parts of the settlement, are precarious wooden constructions. As families grow, houses are expanded vertically and, when possible, also horizontally. A majority of the residents are first- or second-generation immigrants from Colombia. Many have family members or relatives living in the same settlement. Most residents try to increase their incomes by undertaking several jobs simultaneously, such as working as housemaids or informal taxi drivers, selling food and sweets on the street, renting out rooms, or making jewellery from beads and hemp. The settlement has a local committee for addressing community matters, but due to high levels of violence and political factionalism in the area, community engagement and social cohesion strongly declined during the past ten years. You also learn that some residents experienced severe losses during the large-scale disaster that occurred in 1999 (called 'The Vargas Tragedy'). After this event, some residents have stopped investing in their residences with the aim of avoiding losses caused by potential future disasters. Other residents are still very active. Common activities are: (a) removing trash from slopes and water ducts; (b) cutting down trees so that branches cannot fall on people and infrastructure during storms; (c) cementing pathways and slopes in order to facilitate drainage of water and to prevent landslides; (d) storing water in closed containers; (e) during heavy rains, staying alert for cracks appearing in the walls; (f) if cracks appear in the walls, evacuating the house; (g) during heavy rains, sending children to sleep in houses of relatives in more secure areas; and (g) trusting in God to keep them safe.

Task:

1. Identify any hazards, vulnerabilities, used capacities and unused capacities in the settlement.
2. Identify coping strategies for risk reduction and adaptation that could be considered effective and sustainable, and others that should possibly be scaled down or replaced.
3. Identify possible negative effects of 'The Vargas Tragedy' on people's coping strategies.
4. Discuss why it could be an advantage (or a disadvantage) to foster urban risk reduction and adaptation in an area where people have recently experienced a major disaster.

5.9 Guide to further reading

Ayers, J., and Forsyth, T., 2009. Community based adaptation to climate change. *Environment: Science and Policy for Sustainable Development*, 51 (4), 22–31.

Ayers, J. *et al.*, eds., 2012. *Community-based adaptation: scaling it up.* London: Earthscan.

DFID (UK Department for International Development), n.d. *Improving knowledge transfer in urban development.* Output from the Urbanisation Knowledge and Research project: Improving Research Knowledge Technical Transfer. London: DFID.

Dodman, D., and Mitlin, D., 2011. Challenges for community-based adaptation: discovering the potential for transformation. *Journal of International Development*, 25 (5), 640–659.

IFRC (International Federation of the Red Cross), 2004. *World disasters report: from risk to resilience – helping communities cope with crisis.* Geneva: IFRC.

Jabeen, H., Johnson, C., and Allen, A., 2010. Built-in resilience: learning from grassroots coping strategies for climate variability. *Environment and Urbanization*, 22, 415–431.

Satterthwaite, D., ed., 2011b. Community-driven disaster risk reduction and climate change adaptation in urban areas (special issue). *Environment and Urbanization*, 23 (2).

Shaw, R., *et al.*, 2008. *Policy note: indigenous knowledge – disaster risk reduction.* Kyoto: EU, UNISDR, Kyoto University, Seeds.

Wamsler, C., 2007a. Bridging the gaps: stakeholder-based strategies for risk reduction and financing for the urban poor. *Environment and Urbanization*, 19 (1), 115–142 (special issue on 'Reducing risks to cities from climate change and disasters').

Wamsler, C., and Lawson, N., 2012. Complementing institutional with localised strategies for climate change adaptation: a south–north comparison. *Disasters*, 36 (1), 28–53.

World Bank, 2012. *Climate change, disaster risk, and the urban poor: cities building resilience for a changing world.* Washington, DC: The World Bank.

See also recommended readings included at the end of the other book chapters.

5.10 Web resources

CBA Conferences. International Conferences on Community Based Adaptation (CBA) organized by the International Institute for Environment and Development (IIED): http://www.cbaconference.org

Community Monitoring and Preparedness for Natural Disasters Initiative (COMPREND). Launched by the United Nations Development Programme (UNDP) in September 2004, COMPREND is a global initiative which promotes south–south cooperation, the sharing of communities' disaster experiences, and preparedness activities for earthquakes, volcanic eruptions and severe climate-related disasters: http://www.globalwatch.org/ungp/

Inteligencias Colectivas. Inteligencias Colectivas is a horizontal learning system and free database which collects the technical know-how of city dwellers, including information about risk reduction and adaptation planning: http://www.inteligenciascolectivas.org

UNDP Adaptation Learning Mechanism (UNDP-ALM) and its Community-Based Adaptation Programme. The Community-Based Adaptation Programme provides a portfolio of community-level adaptation projects, which were implemented in the programme's participating countries (Bangladesh, Bolivia, Guatemala, Jamaica, Kazakhstan, Morocco, Namibia, Niger, Samoa and Vietnam): http://www.undp-alm.org, http://www.undp-alm.org/projects/spa-community-based-adaptation-project

See also web resources included at the end of the other book chapters.

Part 3
Moving forward

6 Towards sustainable urban risk governance and transformation

Learning objectives

- To compare different theoretical and practical approaches to risk reduction and adaptation planning in cities
- To reflect upon the implications of the gaps identified between different theoretical and practical approaches for improving sustainable urban risk governance and transformation
- To identify ways in which city dwellers' adaptive capacities could be (better) supported by city authorities

Understanding the differences and gaps between the theoretical and the different practical approaches taken to reduce and adapt to disasters and climate change can yield important insights for reforming urban actors' work to more effectively address increasing risk. This chapter summarizes and highlights the key challenges and gaps presented in Chapters 4 and 5, and discusses them together with potential ways in which these could be addressed.

6.1 Theoretical versus practical planning approaches

Chapter 4 focused on city authorities' adaptive capacities by analysing (a) their risk reduction and adaptation measures taken at the household/programme level (Sections 4.1 and 4.3) and (b) their strategies employed for mainstreaming risk reduction and adaptation into urban planning practice (Sections 4.2–4.3). Before continuing to read, ask yourself: are the identified measures and strategies in line with, or different from, the theories and concepts presented in Chapters 2 and 3?

Very few city authorities adequately combine the possible measures on hand and, at the same time, manage to anchor these efforts at institutional levels to achieve sustainable change. The identification and combination of adequate measures and strategies seems to be one of the key challenges. Whilst a range of different measures is increasingly pursued at the programme level to foster urban communities' disaster resilience, they seldom address risk in a comprehensive way (see Chapter 2, Box 2.4), nor are they sufficiently sustained through organizational, interorganizational, financial and legal structures and mechanisms. Moreover, only a few city authorities engage in all strategies that are necessary to achieve the sustainable mainstreaming of risk reduction and adaptation (see Chapter 2, Boxes 2.8–2.10). Such a comprehensive approach is, however, crucial, even for those city authorities that are today well prepared for 'conventional' weather extremes (such as riverine flooding) because the systems and mechanisms associated with these extremes are becoming more and more 'outdated' in a context of climate change (resulting, for instance, in increased flash floods).

Taking a closer look at the existing measures at the programme level (see Tables 4.1–4.5), it can be observed that most efforts focus on addressing the (potentially) most visible and direct disaster impacts on the urban fabric. Examples are the construction of flood defences, retaining walls or disaster-proof bridges and houses. This grey infrastructure approach and its associated 'grey' measures, however, often fail to adequately consider the city–disasters nexus, i.e. the interlinkages between the urban fabric and environmental, socio-cultural, economic, institutional and political aspects that make cities into hotspots of risk (see Chapter 3, Boxes 3.1–3.4). Cities are living systems composed, not only of buildings and infrastructure, but also of more 'invisible' structures and, most essentially, the inhabitants who guide their functioning. Important as it is to address the visible and direct impacts on the urban fabric, the societal consequences also require proper attention, such as increased urban inequality and poverty leading to the need to house and serve a greater number of vulnerable people, disruptions in formal education, loss of livelihoods, long-term health impacts, disputes over access to increasingly scarce resources, conflicts between displaced disaster victims, loss of key persons and workforce and the aggravation of political and economic stressors (see Section 3.3).

The grey measures that are predominantly employed by city authorities and planners often focus exclusively on reducing hazards or physical vulnerability. They are thus seldom exploited for targeting non-physical vulnerabilities that are linked to the urban fabric (see Sections 3.1–3.2) or for supporting preparedness for disaster response and recovery.

Overall, preparedness for response and preparedness for recovery are given little consideration. Consequently, not many cities are well prepared. An example of the lack of consideration given to disaster preparedness is the fact that risk or loss financing is seldom adequately integrated into housing finance mechanisms (i.e. governmental and non-governmental subsidies, micro-credits, supported mutual help or self-help mechanisms), especially in marginal settlements. In addition, most cities are not well prepared for system failures that would, for instance, require better back-up systems and the reduction of dependencies on external services for water, energy or food supply.

Furthermore, there are very few measures that fully take account of citizens' adaptive capacities and their differential risk patterns. To begin with, city dwellers' exposure to hazards, their vulnerability and the associated (used and unused) adaptive capacities are generally not assessed in any detail (see Chapter 2 and Tables 2.4, 2.5 and 5.8). Consequently, city authorities are seldom aware of people's coping strategies, let alone related costs, their underlying patterns of social behaviour and further capacities. Comprehensive risk assessments that include these aspects are, however, crucial for increasing urban communities' disaster resilience because the different factors that contribute to risk can vary strongly within cities, settlements, housing units and even within single households (see Chapter 3 and Box 5.6).

Whilst it can be argued that city authorities increasingly follow a planning process whereby the already observed climate impacts and citizens' local knowledge are given some importance, current approaches are not sufficient to develop long-term strategies for urban risk reduction and adaptation. They hardly consider future disaster and climate change scenarios and often lack participatory and consensus-based approaches. The theoretically sustained understanding that urban resilience can only be achieved when institutional and people's local efforts support and complement each other still contrasts with current practice.

The analysis of city authorities' measures for risk reduction and adaptation further shows a tendency to address 'the usual suspects,' i.e. those hazards and hazard-related aspects that city authorities are already familiar with. As an example, in Sweden most emphasis is given to floods, whilst relatively little consideration is given to the increase in icy streets and slipping accidents due to changes in precipitation and temperatures (Jonsson 2012), or to increasing temperatures which actually are the cause of more deaths than floods. In fact, while only seven fatalities were reported as a result of floods in Sweden during the period 1901–2010, heatwaves are believed to cause hundreds of premature deaths every year (Forsberg 2012; MSB 2012b). Independent of the particular context, many cities place strong emphasis on improving their water management systems, which have been the focus of human adaptation throughout human civilization and settlement (Easterling *et al.* 2004), and thus constitute a well-known issue for many city authorities. To improve cities' water management, both grey and green measures are employed. The green infrastructure approach is often used as a multi-hazard, win–win or low-regret measure (see Box 3.6). Related measures target the linkages between the urban fabric and the urban environment, which is the aspect of the city–disasters nexus that is given most consideration (see Sections 3.1–3.3 and Figure 3.5).

Apart from the concrete measures that are taken 'on the ground', city authorities increasingly employ a wide range of strategies to mainstream risk reduction and adaptation into their urban planning practice. Nevertheless, the implementation of isolated stand-alone measures (such as building dikes, floodwalls or retaining walls) still seems to be more appealing to them than the modification of their day-to-day work to account for increasing disasters and climate change (see Chapter 2, Box 2.8). The latter is in many respects more challenging. It requires an in-depth understanding of the city–disasters nexus, citizens' differential risk patterns and related root causes of risk – knowledge that generally hardly exists (see Chapters 2 and 3). This lack of knowledge is also related to the continuing divide between those professionals who are responsible for urban planning and those responsible for the management of risk, climate change and disasters. Although in recent years both professional communities have made significant advances towards integration, there is evidence that many planning decisions (e.g. for land use) are still not being taken on the basis of all the available risk and climate information.

The analysis of city authorities' mainstreaming strategies shows that most emphasis is generally given to the revision of regulatory and policy frameworks, followed by efforts to improve interinstitutional cooperation to strengthen existing risk governance structures. As regards the latter, actions are increasingly taken to improve interinstitutional communication and to address competition between different organizations, the lack of institutional cohesion and political manoeuvring. In comparison, improved science–policy integration and the involvement of citizens are rarely considered to improve current risk governance structures (see Chapters 2 and 5, Box 2.10 and Table 2.7). In addition, little attention is given to the revision of the implementing body's organizational structures, financial mechanisms and conventional planning tools to ensure the integration of risk reduction and adaptation at programme level. Finally, little consideration is also given to the risk faced by the city authorities themselves, even when their headquarters, other office buildings or staff are located in high-risk areas (see Box 2.9 and Table 2.7).

Section 6.3, entitled 'Bridging the gaps', presents the planning principles that were identified on the basis of the described gaps between the theoretical and practical approaches at both the programme and institutional levels.

6.2 City dwellers' versus city authorities' approaches

Apart from the aspects mentioned in the previous section, further challenges and gaps can be identified when comparing city authorities' work on risk reduction and adaptation (presented in Chapter 4) with people's own efforts to cope with disasters and increasing risk (presented in Chapter 5). Before continuing to read, ask yourself: what are the key differences between city authorities' and urban dwellers' local-level efforts? And what are the resultant barriers to effective risk reduction and adaptation planning?

In sum, there are hardly any cases where the different efforts of city authorities and urban dwellers to reduce and adapt to risk fully support and complement each other. The heatwaves in Chicago in 1995 and New York City in 2011 are examples that illustrate how this leads to counterproductive results:

> [In Chicago] high demand for electricity caused outages, and much of the city lost water as residents opened hydrants to cool off and then fought with city officials trying to close them . . . In New York City, police officers drive through streets using loudspeakers asking people to turn down their air conditioning during the day. The power grid can't handle it.
>
> (Dorell 2012)[1]

In general, the institutional and policy landscape and the operational interventions in place are inadequate in terms of supporting urban dwellers' local risk reduction and adaptation efforts, which is evidenced by across-the-board inadequacies in policies, policy instruments, regulatory frameworks, data recording and sharing, and the mechanisms and structures for interinstitutional cooperation and risk governance. As an example, existing municipal strategies, legislation and mandates often lack a clear definition of both institutional and individual responsibilities for risk reduction and adaptation, and the borders between them. Where such definitions exist, they tend to obstruct (and not support) each other. Owing to legal restrictions on using municipal investment it may, for instance, be impossible to finance adaptation measures on private land, even if they are for the common good (Leonardsen 2012). One of the reasons for the lack of institutional support for localized risk reduction and adaptation efforts is the fact that city authorities seldom assess, and thus rarely know of, urban dwellers' adaptive capacities and their differential patterns of risk (see Section 6.1). There are city authorities that even consider urban dwellers' coping strategies per se as being harmful in terms of interfering with governmental responses.

To make matters worse, citizens are seldom aware of their responsibility to protect themselves from disasters and climate change impacts. Despite this situation, only sparse information is provided to urban dwellers on how best (or better) to adapt. Where such information exists, its dissemination is limited and often does not reach those most at risk. City authorities rarely provide personal counselling or advice for at-risk communities or disaster victims, and there is a lack of formal incentives to take adequate individual action. In low- and middle-income nations, and especially in marginal settlements, NGOs and CBOs often step in to better link city authorities' and people's efforts, with diverse results. In the northern hemisphere, cities strongly support local action taking only for climate change mitigation – but with some positive spin-offs for adaptation, particularly for adapting to water scarcity. The private sector also provides support (for instance by offering specialized products such as commercial floodgates, water-saving devices or pre-seeded bio-mats).

Further analyses of city authorities' interventions show that the assistance on offer has been framed largely from the hierarchical standpoint, whilst people's coping strategies are diverse and are based on different patterns of social behaviour (see Chapter 5 endnotes 6 and 7 for definitions of individualistic, hierarchical, communitarian and fatalistic patterns of behaviour). Consequently, city authorities' assistance often entails the silencing of all but one pattern of social behaviour (see Thompson 2011). This becomes even more questionable given the knowledge that after disaster occurrence at-risk citizens have a tendency to lose trust in hierarchical institutional structures, and often even fear being 'taken for a ride' by national and municipal authorities. Such mistrust in authorities can be rooted in many issues, including delayed or unequal distribution of assistance, governments' neglect of certain areas, or the persistence of some former (colonial) structures (Box 6.1).

As a result of the described situation, even after institutional assistance is provided, citizens usually continue to cope with increasing risk and disasters *just as they did before* – without having obtained better information or arrangements for improving their own efforts. To make matters worse, there are also cases where city authorities' work has dreadfully hampered city dwellers' ways of coping. Examples are institutional measures that provide people with a false sense of security, or retaining walls that have replaced local techniques and cannot be maintained locally. Other examples are urban development projects that deprive people of their livelihoods without providing new alternatives, such as the resettlement of people far away from their income sources (see Box 3.4), or waste management projects which may remove the livelihoods of people who earn their living by searching for recyclable and sellable materials on waste sites (Hamza *et al.* 2012). The provision of social housing that cannot be used by the homeowners as a guarantee for credits (for future risk reduction or recovering after disasters) is another example of how city authorities' work can hamper local coping (Wamsler 2007a). There are further cases where agreements between the government and the insurance industry were deregulated, leaving an increasing number of people at risk without access to disaster insurance (e.g. Wamsler and Lawson 2011). Remains of historical structures where authorities have sought to control rather than serve the population provide another example, described in Box 6.1.

There are also various cases where city authorities' work has been obstructed by city dwellers' own efforts. The statement about the Chicago and New York heatwaves cited at the beginning of this section shows how this can lead to counterproductive results. Another interesting example is described in Box 6.2 where traditional knowledge on weather patterns has been used by the population without giving any consideration to scientific information provided by the authorities. The outcome was a sad total of about 700 fatalities. Yet another example comes from Chile where, after the 2010 earthquake and succeeding tsunami, many people had lost their trust in 'official' information and warnings. As a result, recurring false alarms were issued locally by coastal communities, causing many casualties and deaths due to massive panicked escapes to nearby hills, related traffic accidents and anxiety attacks.[2] A quite different example that shows how local coping can obstruct city authorities' work comes from the capital of El Salvador where people tend to add extra floors to one-storey buildings or build on top of retaining walls that have been constructed by public authorities, leading to decreased structural integrity and consequently increased risk (Wamsler 2007a).

The lack of synergy between city authorities' and people's efforts results in many city dwellers staying dependent on outside help, calling upon governmental and non-governmental organizations for more assistance. In high-income nations, this situation can lead to a growing dependency on the private sector (such as insurance companies), in low- and middle-income

BOX 6.1

Constraints on local authority–citizen action: the case of Georgetown, Guyana

Local authority and citizen-led projects for risk reduction and adaptation can often be far removed from each other. This is especially so in cities where the local government has become highly politicized or where capacity has been drained. In Georgetown, the capital of Guyana, both situations exist.

Georgetown lies at about mean high water and frequently experiences floods, largely because of poorly maintained drainage infrastructure. Georgetown's city authority is built around an elegant structure. Elected neighbourhood-level councils are charged with maintaining the local environment and working with national agencies and community groups to achieve this, but on the ground, drains are clogged with vegetation or household waste, while machinery sits idle and houses flood during heavy rain or high tides. How can this be? Neighbourhoods where communities are organized and flood risk management is a priority for local residents have no more success.

In this example, local action is constrained fundamentally by the relationship between local authorities and local people. Understanding the barriers to local risk reduction requires some appreciation of Guyana's development history as well as more proximate causes: local authority in Guyana, as in many other countries, was first instituted during colonial rule. In this period and until very recently, local government's primary function was to draw intelligence upon and be an agency of the central government for the top-down management of local development. It is only relatively recently that the stated aim of local government has been to serve local interests. Cultural change takes longer. So, even when local groups approach local government and can pay for access to drain-cleaning equipment or offer local labour, these advances are likely to be refused. Local staff and councillors may look upon such acts as embarrassing or threatening. When local councillors are also prominent voices in local civil society, the overlap can stifle local alternative thinking and organizing. In this way, many of the virtues of local government (closeness to the people at risk, shared visions of development and direct accountability) are distorted into incentives for inaction.

By Mark Pelling

BOX 6.2

Local coping strategies: blessing or barrier for authorities' work for risk reduction and adaptation?

In 2000, the Limpopo River Basin in Mozambique experienced very substantial rainfall for many days as a result of unusual cyclone activity. Experts knew that it would result in serious flooding of a magnitude never experienced before. But which source of information would people trust if their lives depended on it – the authorities or their ancestors?

Most communities had no electricity or radio, yet people were usually able to successfully predict floods by observing ants. Ants build their homes underground and when groundwater rises, they leave their nests; thus people would know that the water is rising. On this occasion, the flood came so rapidly that there was no time for the groundwater to rise or for ants to react before the river overflowed. When someone who had heard the experts' prediction drove to a certain village to tell them to evacuate, the local chief asked him, 'Who are you and why should I do what you say? Since the times of my ancestors, floods have only occurred after ants leave their homes. Now the ants are not moving and you come and ask me to leave?' As in most of the Limpopo Valley, many people did not evacuate. About 700 people drowned.

By Mohamed Hamza, Dan Smith and Janani Vivekananda

nations to continued reliance on NGOs. However, although insurance companies often do not provide the type of assistance that would allow urban disaster resilience to be increased (see Section 4.1), NGOs working in the south have developed a range of bottom-up approaches that have succeeded in fostering local risk reduction and adaptation (see Sections 5.5–5.6).

The described challenges between city authorities' and citizens' approaches for risk reduction and adaptation led to the development of several planning principles presented in the following section.

6.3 Bridging the gaps

While the challenges and gaps discussed in Sections 6.1–6.2 are numerous, there are also many opportunities and many positive advances have been made in urban risk reduction and adaptation. City authorities and urban planning certainly provide a powerful and vital platform for supporting risk reduction and adaptation. Not only in theory is there ample scope for potentially beneficial planning interventions that can foster more disaster-resilient cities and lead to sustainable urban transformation. Already in practice, cities have demonstrated a common ability to strengthen their capacities, and this is regardless of their context-specific baseline conditions (see UNISDR 2012c).

The gaps identified between the theoretical and practical approaches to risk reduction and adaptation planning led to the development of potential solutions for how these could be bridged. They are summarized in Boxes 6.3–6.5 in the form of ten planning principles. These planning principles can provide guidance for both scholarly work and city authorities' and planners' daily practice. In the following sections, they are presented in detail and, wherever applicable, illustrative examples are given. In addition, the planning principles are placed in relation to other guidelines available, i.e. the ten 'essentials' of the UNISDR Making Cities Resilient Campaign (UNISDR 2012b: 25)[3] and the five 'success factors' for planning urban adaptation promoted by the European Environmental Agency (EEA 2012: 79).[4]

The first six planning principles (listed in Boxes 6.3–6.4) refer to planning practice at the household/programme level. They can assist urban actors in (a) dealing with increasing risk more holistically and (b) better supporting citizens' adaptive capacity and making it complementary to their own work. In contrast, the following four principles relate to strategies or modifications at institutional levels (see principles 7–10 listed in Box 6.5). Their aim is to assist in the creation of institutional support mechanisms and structures: first, to ensure that the adequate consideration of risk reduction and adaptation at the programme level becomes a standard procedure for urban planning practice and second, to create the institutional capacities necessary to achieve sustainable urban transformation.

The ten planning principles create an 'integrated engagement model for urban risk reduction and adaptation'. If put into practice, it leads to a more holistic and distributed urban risk governance system where institutional and localized efforts can complement each other and top-down and bottom-up approaches are united. It further promotes giving greater weight to the societal aspects of risk reduction and adaptation, and allows a greater degree of flexibility to be incorporated into measures and strategies to deal with uncertainty and to reduce and adapt to increasing risk. Last but not least, it can foster further rapprochement between the separate professional communities of urban planning, risk reduction and adaptation, all together leading to a greater likelihood of achieving sustainable urban transformation.

6.3.1 Dealing with increasing risk more holistically

Planning principle 1. A thorough risk assessment is an indispensable pre-condition for dealing with risk in a holistic fashion and to carry out any adequate alterations to city authorities' and planners' work on the ground (see Section 2.2; see UNISDR 2012b essential 3). Introducing risk assessment methods in planning procedures – for conducting preparatory studies,

---

BOX 6.3

Planning principles for urban risk reduction and adaptation: ten essentials for moving forward (principles 1–4)

I. Dealing with increasing risk more holistically

Planning principle 1: Address root causes of urban risk by designing local activities on the basis of risk assessments comprising thorough analyses of the city–disasters nexus, related urban–rural linkages and citizens' differential risk patterns (including local coping strategies and further adaptive capacities).

Planning principle 2: Address risk holistically in daily planning practice through the consideration and combination of measures for hazard reduction and avoidance, vulnerability reduction, preparedness for response and preparedness for recovery (both in the pre- and post-disaster context).

Planning principle 3: Exploit every type of potential risk reduction and adaptation measure so as to provide, whenever possible, additional benefits and address different contributing risk factors simultaneously (i.e. hazard[s], location-specific vulnerability, deficiencies in both mechanisms and structures for response and recovery).

Planning principle 4: Address risk holistically by adopting a planning approach that encompasses all elements of urban development and counteracts the vicious reinforcing feedback loops of the city–disasters nexus (including physical, environmental, socio-cultural, economic and political/institutional aspects).

designing, implementing, monitoring and evaluating urban developments – is crucial. Contrary to current practice, risk assessment should not only include hazard information from past events and possible climate change scenarios; it should give other risk factors at least the same (or greater) importance. In order to ensure that root causes of urban risk are addressed, risk assessments need to include location-specific analyses of city dwellers' differential risk patterns (see Chapter 2 and Box 5.6), the city–disasters nexus and related urban–rural linkages (see Chapter 3). City dwellers' differential risk patterns embrace urban societies' exposure to multiple hazards, related vulnerability, existing mechanisms and structures to respond and recover from these hazards, individuals' inherent coping strategies, as well as further elements that contribute to their adaptive capacity (see Section 2.1 and Tables 2.4–2.5; see also planning principle 5). The establishment and maintenance of related up-to-date databases is vital. At a city level, special importance needs to be given to regular risk assessments and monitoring of the safety of schools and health facilities (see UNISDR 2012b essentials 3 and 5).

Planning principle 2. Potential risk reduction and adaptation measures can be grouped in four categories: measures for (a) hazard reduction and avoidance, (b) vulnerability reduction, (c) preparedness for response and (d) preparedness for recovery. Knowledge and consideration of these measures can assist during the assessment, planning, design, implementation and monitoring of any urban development programme (see Figure 4.1 and Box 2.4 for definitions). They provide a frame of reference that allows the systematic search for and implementation of (more) adequate local interventions that tackle risk in a holistic fashion – without, for instance, purely focusing on vulnerability reduction or losing sight of emergency and recovery plans (see UNISDR 2012b essentials 9 and 10). In addition, they assist planners in linking climate change adaptation and mitigation because climate change mitigation can here be considered as part of hazard reduction. Whenever possible, all four measures should be combined. Flood-proof sanitation for vulnerability reduction can, for instance, be combined with (a) the provision of back-up systems for disaster response (e.g. accessible public toilets or biodegradable plastic bags as emergency toilets [see Section 6.3.2]), (b) the training of local builders to allow a quicker recovery in case of destruction, and (c) concrete floodwalls or the filling of former latrine holes to decrease flood and landslide hazards respectively.

Planning principle 3. Because the potential four measures for risk reduction and adaptation are directly linked to the different factors that contribute to risk (see Sections 2.1–2.2 and Figure 4.1), they also provide a good basis for identifying win–win solutions, i.e. activities that address several contributing risk factors simultaneously and, if possible, provide further benefits (see EEA 2012 success factor 3). Identifying such win–win solutions allows urban actors to address risk more holistically and is critical in view of (a) the difficult financing of urban risk reduction and adaptation (e.g. when city authorities' investments in urban development have to serve the whole community and not only a few citizens), and (b) the fact that solutions that deliver additional benefits increase urban actors' willingness to accept and implement them (see Section 4.5; EEA 2012 success factor 3). An example is the design of public areas in such a way that they foster social interaction and, at the same time, provide floodable green zones or elevated platforms above the water level for emergency 'accommodation', thus integrating both vulnerability reduction and preparedness for response (see Section 4.1). Public areas, such as parks or 'beach promenades', might also be designed to inhibit housing developments in risk areas, thus reducing citizens' hazard exposure. Another example is fire hydrants that can, if necessary, be used as temporary water sources during heatwaves (to cool off) or during reconstruction endeavours (when other water sources are non-functional) (see Section 4.1). Furthermore, it is possible to plan direct risk reduction and adaptation measures in such a way that they become win–win measures. Flood protection dikes can, for instance, be designed to also serve as bicycle paths or recreation sites. A quite different example of a win–win solution is the integration of risk financing into existing housing financing mechanisms or the expansion of housing financing mechanisms to finance risk reduction, emergency or recovery efforts. Concrete measures are the integration of (micro-)insurance policies into housing (micro-)credits, subsidies and savings schemes,[5] or making access to housing subsidies conditional upon accompanying (micro-)credits being bundled with property disaster insurance and life insurance, thus integrating both vulnerability reduction and preparedness for recovery (for more examples see Wamsler 2007a).

Planning principle 4. If put into practice, the described planning principles 1–3 translate into the adoption of measures that encompass both the pre- and post-disaster context, as well as all elements of urban development, including physical, environmental, socio-cultural, economic and political/institutional aspects (see Section 3.1 and Figure 4.1). Such a holistic approach is vital to counteract the reinforcing vicious feedback loops of the city–disasters nexus (see Section 3.4) and to capitalize on 'common' measures with the aim of creating more flexible and sustainable systems. Common measures such as the protection of critical urban infrastructure and services, ecosystems and natural buffers (also promoted by UNISDR's ten essentials for making cities resilient [UNISDR 2012b essentials 4, 5 and 8]) can often easily be expanded to become more effective. A combined grey and green infrastructure approach,[6] which addresses the linkages between the urban fabric's physical and related environmental features, is a good example of this. Whilst, however, the combination of grey and green measures is relatively well established and promoted in current practice (e.g. EEA 2012 success factors 3 and 4), measures that deal with other factors and interdependencies of the city–disasters nexus are largely missing. A measure that addresses interdependencies between cities' physical and economic features could, for instance, be the revision of the selection criteria established for urban development bidding processes so as to reduce economic vulnerabilities at the local level. The rules and criteria for approval could be revised to include a quota for non-skilled labour to be contracted locally and a 'job–capital' ratio favouring labour-intensive activities that do not affect overall economic efficiency. This could further encourage enterprises to subcontract microenterprises situated in low-income areas, hence also promoting the utilization of local workforces and locally produced materials. An example of a measure that addresses the link between the urban fabric's physical and related social features is the design of buildings or public spaces that provide protection during disasters and, during normal times, can be used to foster social cohesion. Another example is the exchange of rooms, apartments or houses so that the most vulnerable (e.g. the elderly, disabled or families with small children) obtain less risky dwellings (e.g. on the level/side of the building that is less prone to heat, on the upper floor in a flood-prone area, or in the most accessible location, such as on the ground floor, where there is a need to be able to evacuate quickly) (see Box 4.1).

6.3.2 Better supporting city dwellers' adaptive capacity – and making it complementary to city authorities' work

Planning principles 5 and 6. As citizens' adaptive capacity was identified as crucial, although deficient (see Sections 5.1–5.5), it is essential within any urban development to consider supporting selected coping strategies and offering alternative strategies where needed. For this, urban actors need to have at their disposal a framework for viewing and analysing local adaptive capacities in order to build an adequate knowledge base that provides guidance on how to design programme measures that can:

1. Encourage and scale up effective coping strategies (see Section 5.7)
2. Scale down ineffective coping strategies by offering alternative mechanisms (see Section 5.7)
3. Explore existing (but as yet unused) capacities for supporting risk reduction and adaptation (see Table 2.5)

4. Reduce barriers to local coping (see Sections 5.7 and 6.2)
5. Match institutional responses with local people's efforts by offering measures that transcend individualistic, communitarian and hierarchical patterns of social behaviour while easing away from fatalistic approaches (see Chapter 5 endnotes 6 and 7).

Framework for viewing and analysing local adaptive capacities. Urban actors can identify and assess people's coping strategies in relation to their:

- Objectives – i.e. hazard reduction and avoidance, vulnerability reduction, preparedness for response or recovery, and *ad hoc* measures to respond or recover (see Box 2.4 and Sections 5.1–5.4)
- Thematic foci – i.e. physical, environmental, socio-cultural, economic or political/institutional
- Hazard focus – i.e. hazard-specific, non-hazard specific or multi-hazard measures
- Underlying patterns of social relations – i.e. individualistic, communitarian, hierarchical or fatalist (see Chapter 5 endnotes 6 and 7)
- Timing and stages – i.e. pre- or post-disaster, taken a long time or shortly before potential hazard impact (see Table 5.8)
- Awareness – i.e. planned or *ad hoc* measures, deliberate or unintentional (i.e. taken with or without the objective to reduce or adapt to risk)
- Support – i.e. measures that are carried out with or without institutional support by any governmental authorities, aid organizations or other urban institutions (see Sections 4.3–4.4)
- Knowledge transfer – i.e. the information channels used as a base to learn from and to communicate and transfer coping mechanisms to other citizens within and outside the community (see Section 5.6)
- Effectiveness and sustainability – i.e. their long- or short-term increase of capacities (see Section 5.5).

In relation to the last point, coping strategies can be divided into those that can increase capacities for risk reduction and adaptation in the short and/or long terms and those that reduce capacities in the short and/or long terms (see Section 5.7).

In order to be able to foster local adaptive behaviour, not only citizens' individual coping strategies need to be assessed but also, and most importantly, their coping systems, i.e. the combined set of coping strategies used. In contrast to the effectiveness of single coping strategies, the effectiveness of coping systems has a broader scope with direct links to the concept of sustainability (see Box 2.11).

BOX 6.4

Planning principles for urban risk reduction and adaptation: ten essentials for moving forward (principles 5 and 6)

II. Better supporting citizens' adaptive capacity and making it complementary to city authorities' work

Planning principle 5: Design local activities so that they 'build' on citizens' adaptive capacities and related patterns of social behaviour by transcending individualistic, communitarian and hierarchical patterns, supporting and scaling up local coping strategies where adequate, and scaling down and offering alternative solutions where needed.

Planning principle 6: Make sure that institutional and local efforts for risk reduction and adaptation do not hamper but complement each other to foster a more distributed urban risk governance system and possibly provoke systems change for long-term sustainability (see planning principles 9 and 10). Participatory planning approaches are an obvious precondition for complementing institutional with local efforts.

A coping system can be understood to be effective/sustainable if it assists an individual, household or community in reducing their level of risk, while maintaining or enhancing local adaptive capacities both now and in the future (thus not compromising the ability of future generations to meet their own needs). Because the effectiveness of a coping system is strongly determined by its flexibility and inclusiveness, these two aspects have to be analysed in detail (see Section 5.5). Flexibility firstly relates to the number of measures that address each risk factor and thus to the redundancy in the coping system (see Sections 2.1 and 2.7). Secondly, it refers to the diversity of these measures as regards their thematic and hazard foci and their underlying patterns of social behaviour (see aspects listed above). Inclusiveness relates to the use of not just some but all of the four potential risk reduction and adaptation measures to ensure that all types of risk factors are addressed (see the first aspect listed previously [objectives]; Section 2.2 and Box 2.4). As an example, a coping system in which citizens individually deploy only one physical measure to address a specific risk factor does not provide any back-up system and is thus more rigid than a system in which different physical, environmental, socio-cultural, economic and political/institutional measures are combined to address a specific risk factor through individual and joint efforts (see Box 6.3, planning principle 4). In addition, a system is hardly inclusive if, for instance, only physical vulnerability reduction is employed (see Box 6.3, planning principles 2 and 3).

Finally, other adaptive capacities that do not become apparent in people's individual coping strategies and associated coping systems have to be assessed. These include general aspects describing people's living conditions (such as level of income, level of formal education, health, gender, etc.), as well as other aspects that can potentially be exploited for risk reduction and adaptation, such as existence of social networks, certain professional groups, or available resources such as vehicles (see Section 2.2 and Tables 2.4–2.5). Barriers preventing use of these capacities (to date) should, in this context, also be identified.

Supporting and scaling up. Once the existing local efforts and capacities for risk reduction and adaptation are assessed, the possibility of encouraging and scaling up effective coping strategies and community-based approaches can be assessed. Examples for the support or upscaling of existing strategies and approaches include:

•   Upscaling of local methods or local learning mechanisms for risk reduction and adaptation. As an example, in Kenya a group of young people created the *Ushahidi* website (Swahili for 'testimony')[7] with the purpose of mapping the violent riots that broke out in the country after a disputed election in 2007. The system allowed reports to be made via the internet or SMS and the incidents of violence could then be visualized on a map or timeline. Since then, the open-source platform has been supported by different organizations and was subsequently used in a number of countries to coordinate disaster response: during the Haitian earthquake in 2010, the 2011 earthquake and tsunami in Japan, snow removal in New York City and flood management in Australia (Naone 2011). *Ushahidi* is today a US-registered non-profit organization with staff in Africa and the United States (Jeffrey 2011).

•   Establishment of regulatory frameworks that support or ease local risk reduction and adaptation. An example is the prohibition of plastic bags, which supports people's efforts in clearing their neighbourhoods of litter, either as an individual measure or through community 'cleaning days'. Solid waste, and in particular lightweight plastic bags, discarded in the urban environment clog open water channels, drainage systems and roof gutters and create standing waters where mosquitoes can breed and diseases spread. In Mumbai, India, lightweight plastic bags were an important contributing factor to the floods that took over 400 lives in 2005 and are now banned in the city (CBC News 2007). Montgomery County in the United States not only levies a charge for the use of plastic bags to reduce the litter along the Anacostia River, it further uses the collected money for litter clean-up, watershed restoration and storm water management (see Table 4.2).[8]

- Establishment of regulatory frameworks that promote the upscaling of local risk reduction and adaptation. An example is the Code for Sustainable Homes introduced in the UK in 2007 as a voluntary standard, which includes some aspects related to adaptation (Shaw *et al.* 2007).[9] Whilst citizens in the UK often use water butts in domestic gardens to collect rainwater for watering plants, the code encourages the fitting of large underground tanks to new-built homes to collect rainwater for flushing toilets, washing clothes, watering the garden and washing cars. This can reduce the amount of mains water used by the home by 50 per cent.

- Support and upscaling of local risk reduction and adaptation measures by linking them to related institutional mechanisms. Cities' formal and citizens' informal early warning systems can, for instance, be combined to ensure that everybody is informed in a timely fashion (see Box 4.2). As an example, in San Salvador at-risk communities have been supported (with basic technology and training) so that local knowledge on increasing water levels and precipitation can be communicated to the municipality and the national emergency agency, which in turn can better assist the communities (Wamsler and Umaña 2003; see UNISDR 2012b essential 9).

- Linking community-based risk reduction and adaptation with city-wide responses. An example is so-called Shared Learning Dialogues (SLDs), a participatory planning process for mutual learning and problem-solving in complex situations, characterized by iterative, transparent discussions with local community members, government agencies, civil society organizations, research centres and other technical agencies designed to facilitate mutual learning and joint problem-solving (ISET 2010; see UNISDR 2012b essential 1).[10]

- Assistance for or insurance of local savings schemes. The support of local emergency funds can, for instance, allow communities to access grants and loans for risk reduction and adaptation at favourable interest rates from (micro-)finance institutions, local-level authorities, national governments or aid organizations (Wamsler 2007a).

- Offer or maintenance of land to support local risk reduction and adaptation. In the cities of Malmö and Lund, Sweden, land is offered for gardening and agriculture (so-called urban agriculture), which supports people's own efforts to deal with high temperatures or eventual restrictions in food supply (see EEA 2012 success factor 3).

• Creation of foundations to guarantee insurance coverage. City authorities can 'link up' with insurance companies with the aim of guaranteeing insurance coverage, whilst at the same time offering at-risk citizens help with risk reduction and adaptation, thus reducing human suffering and high (post-disaster) costs for the society.[11]

• Recognition and support of individuals, groups, communities and local organizations that demonstrate leadership and innovation in adapting to and preparing for the impacts of climate change. A concrete example is Australia's National Adaptation Research Facility and its Climate Adaptation Champions programme.[12]

Another way to support (or scale up) local efforts is the creation of incentives and improved knowledge transfer for risk reduction and adaptation, including private counselling, guidance, dissemination of information and better communication in a form that is understandable for the different users (Figure 6.1; see EEA 2012 success factor 1). Concrete examples are incentives for homeowners to invest in reducing the risks they face,[13] education and training programmes on disaster risk reduction and adaptation, and readily available public information and plans for improving citizens' disaster resilience (see UNISDR 2012b essentials 2, 3 and 7). Whilst a one-way and top-down knowledge transfer is not sufficient to achieve sustainable change, it is an important step towards improving current mechanisms and structures. An interesting example showing how people can be incentivized to take adaptation measures is the New York City 'CoolRoofs' project, carried out in collaboration between the Department of Buildings and the City Service. It facilitates the cooling of New York City's rooftops by financially supporting individuals, voluntary groups and civic associations to apply a white, reflective surface to rooftops of residential buildings (Figure 6.2). This activity minimizes heat stress, reduces cooling expenses, cuts energy usage and lowers greenhouse gas emissions. Coating all eligible dark rooftops in New York City could result in up to a one-degree reduction of the city's air temperature. In addition, it can be seen as a boost for physical activity, social cohesion and green thinking (NYCDOB 2011). Another type of incentive is conditional cash transfers for low-income groups, on the condition that certain risk reduction measures are implemented (see Section 4.4).

Scaling down by offering alternatives. Apart from supporting effective coping strategies, ineffective strategies can be scaled down and alternatives

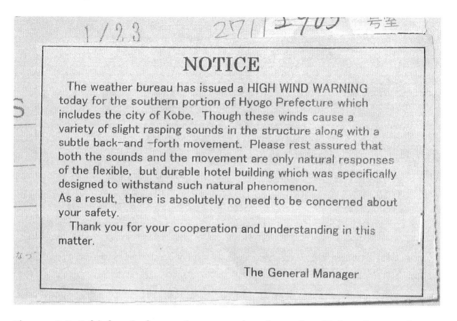

NOTICE

The weather bureau has issued a HIGH WIND WARNING today for the southern portion of Hyogo Prefecture which includes the city of Kobe. Though these winds cause a variety of slight rasping sounds in the structure along with a subtle back-and-forth movement. Please rest assured that both the sounds and the movement are only natural responses of the flexible, but durable hotel building which was specifically designed to withstand such natural phenomenon. As a result, there is absolutely no need to be concerned about your safety.

Thank you for your cooperation and understanding in this matter.

The General Manager

Figure 6.1 A high wind warning on a hotel receipt, Kobe, Japan. Source: Christine Wamsler.

offered. The result of new structures should be that the population and/or the environment are better off than under the current situation. If they only replace what is there, it is not a great help.[14] In addition, a seemingly ineffective coping strategy might (although being short-term in itself) be an important and irreplaceable part of a long-term solution by complementing other strategies. The whole coping system thus needs to be carefully considered before any actions are taken (see Section 5.7). An example of scaling down by offering better alternatives is the improved access to formal insurance systems or the offer of formal money-transfer systems, which may be more efficiently delivered than private transfers (see Section 4.2, case of Manizales). A very different example relates to the so-called 'flying toilets', i.e. people relieving themselves into plastic bags and tossing them out the window during water scarcity (see Section 5.5). The commercialized product PeePoo has been developed to offer a better, although temporary, solution. It is a biodegradable bag which can turn human waste into fertilizer and eliminate infectious pathogens in the process, thereby preventing the spreading of disease and pollution of the environment, whilst maintaining people's level of security (because it can be used at home). The bag is also used in disaster response in order to prevent disease spreading in the aftermath of disasters.[15]

Figure 6.2 New Yorkers painting rooftops of residential buildings white to reduce indoor temperatures and heat island effect. Source: BBC/Olatz Arrieta.

Reducing barriers for local coping. In some cases, avoiding the creation of barriers to effective local coping or reducing existing barriers is already an important step towards improving urban risk reduction and adaptation. Examples of related action taking include:

- Not providing people with a false sense of security, for instance by combining urban organizations' physical risk reduction measures with awareness raising and training
- Not imposing measures that cannot be maintained in the future, such as retaining walls which replace local techniques and cannot be maintained locally
- Allowing homeowners to use assisted social housing as a guarantee when applying for credits to improve their disaster resilience
- Not implementing urban development programmes that deprive people of their livelihoods without providing new alternatives

- Assisting people in informal settlements in such a way that they can also access formal mechanisms for risk reduction and adaptation (such as credits or micro-insurance)[16]
- Establishing conditions that allow people to take adequate and efficient *ad hoc* measures (such as providing easy access to cool/warm places or being able to work from home if needed).

Exploring existing (but unused) capacities for supporting risk reduction and adaptation. Apart from urban communities' 'visible' adaptive capacities in the form of coping strategies, they also possess other capacities that might be explored to foster urban risk reduction and adaptation (see Table 2.5). An illustrative example comes from Rio de Janeiro's *favelas*, where kite flying is a popular pastime for children and where many people possess a mobile camera phone. These capacities were identified and exploited in a UNICEF project (Caparelli *et al.* 2012). Youngsters were trained in mapping socio-environmental risks and vulnerabilities using geographic information system (GIS) technology. Risks were documented with mobile cameras, whereupon photos and geographical coordinates were uploaded to a mapping platform. The project involved an innovative way of taking pictures from above using camera phones tied to kites, providing an (affordable) alternative to outdated maps and satellite photos. Other capacities that could be exploited are, for instance, the existence of local builders, who can be trained in disaster-resistant construction, or well-established local organizations and associations, which can become engaged in risk reduction and adaptation (such as environmental action groups, eco-groups or religious associations) (see Table 2.5).

Matching institutional with localized efforts. The previously mentioned framework for analysing coping strategies and coping systems also allows searching for multi-dimensional approaches that match the different patterns of social behaviour which underlie people's ways of coping (see Box 6.4, planning principle 5). Such multi-dimensional approaches or measures are commonly termed 'clumsy' solutions. They are an alternative to common approaches that silence all but one of the four patterns of social behaviour. Clumsy solutions are often 'initially invisible options that emerge from the messy and argumentative process in which each voice is able to make itself heard and is also responsive to the others' (Thompson 2011: 12). It enables 'each set of actors to get more of what they want, and less of what they do not want, than they would have received if they had silenced the other voices and imposed their own elegant solution' (Thompson 2011: 13). A demonstration of a clumsy solution is the *Bhattedanda Milkway*, a cable system that moves metal carriers containing fresh milk across the Kathmandu Valley of

Nepal, one of the most disaster-prone areas in the world. As described in Thompson (2011) and Thompson *et al.* (2007), given the rugged and disaster-prone environment, this two-mile link is an efficient way of getting milk to market in the capital city before it spoils. The milkway is economic growth achieved by means of a carefully planned intervention that would never have happened 'autonomously'. Hence, those speaking with the hierarchical voice have certainly achieved much of what they wanted. Individualistic actors, the farmers who had their incomes immediately increased by 30 per cent, have also come out well, especially when taking into account the host of innovations that have shifted them away from subsistence agriculture into a form of market gardening (because farmers realized that they could also use the milkway to export perishable farm products to the market and diversified into tomatoes, green vegetables and beans). Increased income diversification has further improved people's way of coping with the many hazards they are exposed to (PDC 2005). And those who speak with the egalitarian or communitarian voice have secured the conservation of the forest (because villagers no longer have to boil their milk down into *khuwa*, a longer-lasting but less valuable condensed milk – thus stopping the cutting of 400 tonnes of firewood per year), along with a convincing demonstration of how climate change mitigation and adaptation can go hand in hand. In fact, the risk of both landslides and floods has been reduced (PDC 2005) and the transition from fossil fuels to renewables has been effected (because the villagers converted the power source of the milkway from imported diesel to local hydro-electricity). Moreover, it is the 'poorest of the poor' who have benefited most because their economic vulnerability has been reduced. Improved revenues allow the village milkway committee to provide loans to local merchants at interest rates far below those charged by banks and moneylenders.

This example shows that it is important and possible to take advantage of existing local efforts and heterogeneity to design measures that appeal to the various perspectives and efforts of the related stakeholders instead of offering uniform and unidimensional solutions. But this is only the start. People do not just have strategies – under certain conditions they can move from the underlying social pattern of one strategy to another pattern (Thompson and Wildavsky 1986). City authorities, aid organizations and other urban actors can help or hinder such transitions. Therefore, when selecting programme measures, it is important to be sensitive to people's behaviour in different communities so that the social patterns can be either matched or, where needed, channelled. Introducing systems of mutual rights, accountability and community organization to support a shift in social patterns could, for instance, be an important step in overcoming asymmetric patterns of risk (i.e. the unequal distribution of the level of disaster risk experienced by city dwellers living close to and within a specific area). This, however, needs to be followed up over time (i.e. after programme implementation) and needs to be complemented with further strategies, such as trying to ease people away from fatalism and offering solutions for improved individual coping.

Complementing institutional with localized efforts. All together, the framework for viewing and analysing local coping capacities described previously is essential to ensure that institutional and localized efforts for risk reduction and adaptation can complement each other and foster a more distributed urban risk governance system (see Box 6.4, planning principle 6). Increases in efficiency, flexibility and cost effectiveness are some of the immediate outcomes. A simple example comes from the city of Copenhagen, Denmark where a cost-benefit analysis has shown that flood risk could best be reduced by a combination of institutional efforts (to upgrade sewers) and citizens' use of backflow prevention devices (also called non-return valves) to block water pipes when water rises (Hasling 2012). Complementary and harmonized efforts for risk reduction and adaptation have the potential to boost incremental improvements in existing approaches, which can help maintain systems function, as well as provoke systems change for long-term sustainability. Participatory approaches are an obvious precondition for complementing institutional with local efforts. It is indeed not enough to put citizens' needs at the centre when dealing with risk in the context of development, response or recovery; citizens must be given an active role in urban risk reduction and adaptation (see UNISDR 2012b essentials 1, 3, 9 and 10). There are broad experiences with participatory risk reduction and adaptation from which to learn, especially in southern cities (see Section 4.4).

6.3.3 Creating institutional support mechanisms and structures (for urban risk reduction and adaptation)

The first six planning principles described in the foregoing sections (6.3.1 and 6.3.2) refer to urban actors' daily planning practice at the local programme level. The additional four principles, presented in this section, relate to strategies and modifications at institutional levels (Box 6.5). They can assist in the creation of institutional support mechanisms and structures: first, to ensure that the adequate consideration of risk reduction and adaptation at local programme level becomes a standard procedure in urban planning practice and second, to create the local and institutional capacities necessary to achieve sustainable disaster resilience and urban transformation.

BOX 6.5

Planning principles for urban risk reduction and adaptation: ten essentials for moving forward (principles 7–10)

III. Creating institutional support mechanisms and structures

Planning principle 7: Institutionalize risk reduction and adaptation through a complete set of possible mainstreaming strategies to ensure that all relevant aspects for achieving comprehensive integration are addressed.

Planning principle 8 (integral part of 7): Make sure that the organization revises all, and not only some, of the key aspects required to ensure that risk reduction and adaptation become a standard procedure in daily planning practice, ranging from regulatory frameworks and policies to financial and human resources, working structures, organizational mandates, responsibilities and planning tools.

Planning principle 9 (integral part of 7): Ensure improved interorganizational collaboration with other urban stakeholders to harmonize efforts, promote improved science–policy integration and jointly foster a (more) distributed urban risk governance system (including governmental, private, civil society, research and training institutions working in the fields of development, response and recovery, as well as citizens [see planning principle 6]).

Planning principle 10: Combine and complement measures for risk reduction and adaptation with mainstreaming strategies (at the local programme and institutional levels) to support flexible and inclusive risk reduction and adaptation approaches that maintain systems functions and, if necessary, provoke systems change for long-term sustainability.

Planning principle 7. In order to create an institutional support mechanism for urban risk reduction and adaptation, mainstreaming needs to take place in all sectors and at all levels of society (see EEA 2012 success factor 2). Urban organizations thus need to employ the whole range of existing mainstreaming strategies. They are described in detail in Boxes 2.8–2.10 and Table 2.7. The combination of the different strategies is aimed at achieving a profound and sustainable change in the multilevel institutional framework of urban actors, related policies, decision-taking structures and actions, which ultimately results in the delivery of risk reduction and adaptation measures on the one hand and the building of adaptive capacity on the other. The adequate combination and prioritization of these mainstreaming strategies is location-specific and depends on differences in existing hazards, vulnerabilities and related capacities (see Section 2.4).

Planning principle 8. In order to ensure that daily planning practice accounts 'by default' for increasing disasters and climate change, it is crucial that urban organizations revise all, and not only some, of the influencing factors: their regulatory frameworks, policies, financial and human resources, internal working structures, institutional mandates, responsibilities, planning tools, as well as the mechanisms and structures for interinstitutional collaboration at municipal and national levels (see Boxes 2.9–2.10 and Tables 2.7, 4.7 and 4.8). The latter is discussed under planning principle 9. All aspects are briefly presented in the following text to highlight some potential measures. Additional information can be found in Sections 2.4, 2.7, 4.2 and 4.5.

Regulatory and policy frameworks. An important element in ensuring the mainstreaming of risk reduction and adaptation into urban planning practice is the creation of appropriate regulatory and policy frameworks that can (a) ensure a coherent approach, (b) facilitate cooperation between stakeholders across sectors and levels, and (c) support adaptive behaviour (and prevent maladaptive behaviour) of citizens, the private sector, governmental agencies and other urban stakeholders (see UNISDR 2012b essentials 1 and 2; EEA 2012 success factor 2). Whether or not the 'multitude' of small-scale private and public decisions do or do not contribute to increasing cities' resilience depends, in fact, on what city authorities encourage, support and prevent (as well as how well their own work is backed up by municipal, national, regional and international frameworks). At a municipal level, this involves the revision of existing regulations and policies and possibly the creation of new ones. Adequate frameworks for land-use management and building standards are crucial in this context (see Sections 4.1–4.2; UNISDR 2012b essential 6), as well as legal frames of legislation that allow institutional and individual capacities to support each other (see Section 5.7).

Financial resources. The many competing interests and demands on the resources of urban organizations at different levels certainly influence political commitment for risk reduction and adaptation (see Table 4.10). Promoting risk reduction and adaptation as cross-cutting topics, which should – as a matter of good practice – be incorporated into urban planning, may prevent them from being viewed as an additional area of investment that is directly competing for funding with other fields. Although there are many risk reduction and adaptation measures that can be implemented at low cost, a certain and tailored resource base is vital for pushing forward the mainstreaming of risk reduction and adaptation into urban planning practice (see EEA 2012 success factor 5). Cities can tap into different sources of financial support, including (a) mainstreamed municipal budgets (for urban development planning), (b) budgets for risk reduction and adaptation, (c) budgets for disaster response and recovery, (d) partnerships with local and international organizations, (e) technical assistance, and (f) the private sector. In addition, city authorities can provide incentives for others to invest in reducing the risks they face, including citizens, the private and public sectors (see planning principle 5; UNISDR 2012b essential 2). A quite different, but equally important aspect is to make existing organizational budget lines 'climate-proof' so that they are not impacted by increasing disasters and climate change (e.g. resulting in money for urban development being disbursed on disaster response and recovery, in general, or on repairs of former urban development programmes work, in particular).

Working structures, institutional mandates, responsibilities and planning tools. It is important that city authorities, planners and other urban actors are aware of their role and responsibilities in disaster risk reduction and adaptation (see UNISDR 2012b essential 1; EEA 2012 success factor 1). Institutional mandates and staff contracts might need to be revised to explicitly reflect this responsibility (see Table 4.7 and Sections 4.2 and 4.5). In addition, and in order to push forward related action taking, the temporary creation of specialized departments or focal points for mainstreaming risk reduction and adaptation, together with other types of decision-support systems for organizational learning, might be important to assist planners with practical information in relation to:

- Potential physical measures that can be used as an entry point to address (step-by-step) interrelated non-physical risk factors of the city–disasters nexus (and their illustration with concrete project examples)[17]
- Potential measures to (a) avoid or reduce hazards, (b) minimize location-specific vulnerabilities, (c) improve mechanisms and structures for response, and (d) recovery (see Box 2.4 and Section 4.1)
- Potential mainstreaming strategies and step-by-step guidance for their implementation (see Boxes 2.8–2.10, Table 2.7 and Section 4.2)
- Synergies and conflicts that may result from a particular combination of adaptation measures and mainstreaming strategies
- Good practice, research and other relevant background documents from high-, middle- and low-income nations to facilitate knowledge transfer between cities located in both the northern and the southern hemisphere

(including north–south, south–north, north–north and south–south transfer of knowledge) (see Sections 4.4 and 5.6)
- Ways to improve science–policy integration by creating better opportunities for researchers and policymakers to exchange information and actively collaborate
- Planning tools and methods for achieving sustainable urban risk governance involving urban authorities, civil society, the private sector and citizens at risk (see Sections 2.5 and 4.3)

As regards the last, the revision and improvement of city authorities' conventional planning tools (such as environmental assessments or tools for construction control and monitoring) is vital to ensure increasing resilience in cities, as is the adoption of additional tools where needed. This might include the incorporation of risk assessment or methods for participatory adaptation planning (see Boxes 6.3–6.4, planning principles 1 and 6).

Planning principle 9. The ninth planning principle is an inherent part of planning principle 7. It highlights the importance of ensuring improved interorganizational collaboration with other urban stakeholders in order to harmonize efforts, improve science–policy integration and jointly foster a more distributed urban risk governance system (see Section 2.4, Box 2.10 and Tables 2.7 and 4.8). This includes relevant stakeholders from governmental, private, civil society, research and training institutions working in the fields of development, response and recovery, as well as citizens (see Section 2.3). To harmonize different efforts and create a more distributed urban risk governance system, it is critical to establish the interests, roles and responsibilities of the key actors, negotiate conflicting interests and identify the adequate media and strategies for communication (see Sections 4.4 and 5.6). In this context, creative and innovative communication and platforms are vital to enhance learning and action (such as mobile phone and streaming technologies, urban games or urban living labs). Public–private partnerships can also be an important factor in fostering more distributed urban risk governance (see Table 4.8). 'Adopt-a-Light' is the name of a public–private partnership in Nairobi, Kenya, which ensures the streets and slums are lit up at night to enhance road security and reduce crime, factors that are also of relevance for risk reduction and adaptation. The initiative emphasizes the fact that all citizens, including those who live in informal settlements, have the right to safety (UNHABITAT 2010b). The previously mentioned foundation and collaboration between city authorities and insurance companies is another example of a possible public–private partnership (Section 6.3.2). Notably, due to the complex and dynamic nature of the city–disasters nexus and related urban–rural linkages, improvements in urban risk governance need to go beyond the city level, including local, regional, national and international approaches (EEA 2012 success factor 2). As regards improved science–policy integration, there are many different ways to better link scientific knowledge with local policy decision-making. For example, it can be promoted through 'translating' scientific outcomes into policy recommendations; training and dissemination of research outcomes; seminars on adaptation and risk reduction focused on the science–policy interface; the creation of opportunities for researchers, practitioners and policymakers to exchange information; the inclusion of policymakers and practitioners in research projects; and increasing the involvement of local government in universities (see Figure 2.8, Box 2.10 and Table 4.8).

Planning principle 10. The described ten planning principles create together an 'integrated engagement model'. If put into practice, it leads to a more holistic and distributed urban risk governance system and develops the capacities necessary to achieve urban resilience and a sustainable transformation of cities, which can also deal with uncertainties (see Section 6.4 on sustainable transformation). The promoted measures for risk reduction and adaptation at the programme level and the mainstreaming strategies at institutional levels should therefore not be seen in separation. In fact, the local level measures and the different levels and aspects of mainstreaming (addressed through the various strategies) need to influence each other (see Figure 2.8, Table 2.7 and Boxes 2.8–2.10). The involvement of universities in urban development programmes for the training of informal builders in disaster-resilient construction techniques is a simple example of how to reduce vulnerabilities on the ground, whilst also addressing better science–policy integration that supports the mainstreaming of risk reduction and adaptation at institutional levels. The case of Manizales, described at the end of Section 4.2, provides a comprehensive example to illustrate how the different measures for risk reduction and adaptation and related mainstreaming strategies can be combined to support each other.

6.4 From disaster resilience to sustainable urban transformation

Comprehensive urban risk reduction and adaptation, as promoted by the 'integrated engagement model' and its ten planning principles presented in Boxes 6.3–6.5, is much more than identifying and acting on specific disaster and climate change impacts. It looks to the performance of each city's complex and interconnected physical, environmental, social, economic and political/institutional features and systems that shape urban risk (see Chapter 3), and also includes the development of institutional and individual capacities to withstand or deal with expected and unexpected changes (see Chapters 4 and 5). It aims to shape risk reduction and adaptation processes (rather than a blueprint of specific actions), which has the ultimate goal of creating a culture of resilience.

Disaster resilience is here understood as a process and attribute that can be assessed by (a) the extent to which city authorities and other urban actors are successful in (improving capacities for) reducing current and future hazards, location-specific vulnerability and the deficiencies in response and recovery mechanisms and structures (with particular attention to how this serves at-risk citizens), and (b) the degree to which they succeed in mainstreaming risk reduction and adaptation at institutional and interinstitutional levels (see Sections 2.2 and 2.4, respectively). The visible result is urban actors that are (more or less) able to reduce and adapt to evolving and changing risk in a flexible, dynamic and effective manner.

In line with this, urban risk reduction and adaptation include not only the improvement of risk reduction approaches to maintain systems functions, but also the transformation of systems for achieving long-term sustainability (see planning principle 10). In fact, comprehensive risk reduction and adaptation, as promoted by the 'integrated engagement model', incorporates the notion of 'transformation' (see Section 2.1 and Box 2.3) and 'sustainability' (see Box 2.11), ensuring that related action taking contributes to more sustainable development pathways. It provides the capacities to transform fundamental attributes of urban systems (e.g. governance regimes, knowledge production systems and urban lifestyles) to increase the institutional and individual abilities to shape the physical, environmental, socio-cultural, economic and political/institutional conditions of the future (by linking them to the root causes of urban risk). Such an approach is crucial because the magnitude of changes, including greater frequency and severity of extreme events, trends in temperature and shifts in seasonality, can mean that an incremental approach may become no longer viable. In addition, incremental approaches might support development pathways that have created risk in the first place. As long as the root causes that nurture (increasing) risk are not addressed, incremental improvements that build on existing structures (instead of altering them) are bound to reproduce risk.

The 'integrated engagement model' thus allows the (development of a) combination of relatively small changes to existing approaches and scaling-up efforts, alongside transformative initiatives. It can further be linked to the notion of human security in the sense that it supports capacity development to meet short-term needs and rights as well as longer-term needs and fulfilment.[18] It advocates in fact the creation of an enabling environment for the control of risk and promotion of sustainability (including the disaster resilience of appropriate elements) in which citizens can take an active stake.

6.5 Summary – for action taking

Many gaps and challenges can be identified when comparing different theoretical and practical approaches to risk reduction and adaptation planning in cities, which can yield important insights for increasing knowledge and organizational learning for more effectively addressing disasters and climate change. At the same time, many opportunities and positive advances have occurred. City authorities and urban planning provide a powerful and vital platform for supporting risk reduction and adaptation. Not only in theory is there ample scope for potentially beneficial planning interventions that can foster more disaster-resilient cities and ultimately lead to sustainable urban transformation. In practice, city authorities have already demonstrated a common ability to strengthen their capacities, and this is regardless of their context-specific baseline conditions.

The gaps identified between the theoretical and practical approaches to risk reduction and adaptation planning led to the development of potential solutions to how these could be bridged. They bring together several of the action points presented in earlier chapters (see summary sections in Chapters 2–5) which were complemented with additional findings. The result is summarized in the form of ten planning principles presented in Boxes 6.3–6.5.

The ten planning principles create an 'integrated engagement model' for urban risk reduction and adaptation that, if put into practice, leads to a more holistic and distributed urban risk governance system where institutional and citizens' localized efforts can complement each other and top-down and bottom-up approaches are united. Its integrated approach can foster further rapprochement between the (often still separate) professional communities of planning, risk reduction and adaptation. In addition, it gives more weight to the societal aspects of risk and allows incorporation of a greater degree of flexibility into measures and strategies to reduce and adapt to increasing risk and deal with related uncertainty. Altogether, the ten planning principles not only boost incremental improvements in existing approaches that help maintain functioning systems, but also provoke transformative changes, i.e. systems change for long-term sustainability.

Uncertainties in climate change projections and the complexity of urban risk reduction and adaptation should not be an excuse for not addressing increasing risk in a systematic and comprehensive way. The 'integrated engagement model' is flexible but concrete enough to respond to issues of uncertainty and complexity. Universally valid examples of the proposed planning principles do not, however, exist. It is not an 'off-the-shelf template' or a step-by-step approach, but a kind of mental map of what a risk reduction and adaptation process entails. This is because the identification and design of concrete measures and strategies has to be location-specific, i.e. they have to respond to and match the local characteristics, needs and capacities of both the affected population and the institutions that (should) support them. Depending on the location, the urban fabric needs to be designed, for instance, to withstand either the effects of extreme heat, strong winds, humidity, threat of floods, landslides or earthquakes, or a combination of these. In addition, it has to respond to constantly changing risk patterns, influenced by issues such as yearly seasons, rapid urbanization, differences in pre- and post-disaster conditions and, not least, increasing climatic hazards.

6.6 Test yourself – or others

The following questions can be used to test yourself or others. The answer to most questions can be found explicitly in Chapter 6.

1. What are the key differences between current theoretical and practical approaches to risk reduction and adaptation planning in cities?
2. What are the key differences between citizens' and city authorities' efforts taken to reduce and adapt to increasing risk caused by disasters and climate change?
3. How can citizens' and city authorities' risk reduction and adaptation efforts become more complementary?
4. How can urban organizations support citizens' adaptive capacities? Please provide concrete examples/measures.
5. What are the aspects that urban organizations need to consider in order to address existing risk (more) holistically?
6. What are the key aspects that have to be revised at an institutional level to make sure that (the consideration or mainstreaming of) risk reduction and adaptation at local programme level becomes a standard procedure in urban planning practice?
7. What does the notion of 'urban transformation' entail? How is it related to today's understanding of urban risk reduction and adaptation?

Case studies might also be a good basis for assessing the knowledge presented in this chapter.

 Test yourself scenario 6.1: sustainable urban transformation. Take the description of a specific urban at-risk area (e.g. Box 6.1 and related references).

Task. Discuss or answer the following issues/questions:

1. What types of short-term solutions/measures could be proposed for this particular case?
2. What kind of risk reduction and adaptation measures might be needed to achieve long-term transformation? Note that the transformation concept implies that, at some point, there needs to be a switch in focus from narrower, short-term solutions (such as building higher flood defences) to broader, long-term approaches.
3. Take one of the measures you have proposed (short- or long-term) and analyse how it could be designed to include different patterns of social behaviour, enabling each set of actors to get more of what they want and less of what they do not want.

Test yourself scenario 6.2: 'clumsy' solutions to match institutional with localized efforts. Take the description of a specific urban at-risk area, which includes a description of different measures taken for risk reduction and adaptation, for instance case studies from the online tool entitled 'Learning to Tackle Climate Change' (see Section 6.8) or the following case described by Hausler (2010): The Bhuj earthquake in Gujarat, India took more than 20,000 lives and destroyed over 215,000 houses. Owners of destroyed houses could either rebuild their own homes with cash assistance (homeowner-driven approach) or move into a house built by a non-profit or government organization (donor-driven approach). Approximately 77 per cent of homeowners chose to build the house themselves, and were later reported as being the most satisfied with their new homes. In the homeowner-driven approach, the new houses were built on the original sites, and the owners chose the floor plan and building materials. They could build whatever type of house they wanted. They could also build a larger house if they could afford it, provided they reinforced it properly and followed official reconstruction guidelines. Most homeowners did not actually build a house themselves; instead, they hired local builders and took advantage of technical assistance provided by government-trained engineers. The government provided funds in instalments; homeowners had to comply with the reconstruction guidelines to receive the next instalment. Because they hired the builder and oversaw the construction themselves, the homeowners were more confident that their house could keep their family safe. In the donor-driven approach, houses were built en masse by contractors, usually at relocation sites. The homeowners had little, if any, role in the design and construction. Since the houses were built primarily with donor funds, they were not subject to inspections like the other houses. Still, many of the non-profits followed or exceeded the reconstruction guidelines in order to maintain their reputation and gain the homeowners' trust, and houses built by the local government complied with all earthquake-resistant building norms. But many people in the donor-driven houses were not satisfied with them. Some houses were never occupied, some abandoned their new homes or modified them in ways that were no longer earthquake resistant. Many had architectural or design features that were not appropriate for the climate or the culture. Many had low ceilings, making them unbearably hot. Toilets were not used because they were built inside houses although homeowners preferred them outside. House entrances led to the street rather than to an enclosed courtyard.

Task:

1. Assess how city authorities have considered local knowledge and people's capacities.
2. Analyse the different institutional measures taken as regards their relation to existing patterns of social behaviour (i.e. individualistic, communitarian, hierarchical and fatalistic).
3. On this basis, discuss why some measures where more (or less) successful than others.

6.7 Guide to further reading

EEA (European Environment Agency), 2012. *Urban adaptation to climate change in Europe: challenges and opportunities for cities together with supportive national and European policies* (No. 2). Copenhagen: EEA.

Jha, A., Bloch, R., and Lamond, J., 2012. *Cities and flooding: a guide to integrated urban flood risk management for the 21st century.* Washington, DC: The World Bank.

Loftus, A.-C., *et al.*, 2011. *Adapting urban water systems to climate change: a handbook for decision makers at the local level.* Freiburg: ICLEI European Secretariat.

Shaw, R., Colley, M., and Connell, R., 2007. *Climate change adaptation by design: a guide for sustainable communities.* London: Town and Country Planning Association (TCPA).

UNISDR, 2012b. *How to make cities more resilient: a handbook for local government leaders. A contribution to the global campaign 2010–2015 'Making cities resilient – my city is getting ready!'* Geneva: United Nations.

Wamsler, C., and Brink, E., 2014. Moving beyond short-term coping and adaptation. *Environment and Urbanization*, April 2014, special issue entitled 'Towards resilience and transformation for cities' (forthcoming).

World Bank, 2013. *Guide to climate change adaptation in cities.* Washington, DC: The World Bank.

See also recommended readings included at the end of the other book chapters.

6.8 Web resources

Adaptation Navigator. The Climate Change Adaptation Navigator is a web-based guidance framework designed to assist administrators and decision-makers in local government and other institutions (in Victoria, Australia) to adapt to the impacts of climate change: http://www.adaptation-navigator.org.au

European Climate Adaptation Platform. An initiative of the European Commission helping users to access and share information on climate change and adaptation (e.g. on expected climate change in Europe, national and transnational adaptation strategies, case studies and potential adaptation tools): http://climate-adapt.eea.europa.eu

KlimaExWoSt, Stadtklimalotse. A German online decision-support system to support city authorities with practical know-how on climate change adaptation planning: http://www.stadtklimalotse.net

The Learning Hub for Low Carbon Climate Resilient Development. An interactive online tool aimed at supporting professionals (independent of country, context or personal expertise) to reflect on different aspects of climate change and development in relation to their own work. The Hub is structured around four themed learning cycles, of which one is 'Approaches to Planning for Climate Change': http://www.ids.ac.uk/idsproject/learning-hub

UK Climate Impacts Programme (UKCIP). A British online decision-support system for climate change adaptation to help organizations consider climate risk and plan to adapt: http://www.ukcip.org.uk

See also web resources included at the end of the other book chapters.

7 Concluding remarks

Increasing disasters and climate change pose a serious threat to sustainable development, placing many cities at risk. But how could cities better withstand, counteract or overcome both climate- and non-climate-related hazards? How could cities regain their historical function as places where citizens can find safety and protection?

Identifying adequate solutions is an urgent task for city authorities and other urban actors around the world, whilst knowledge and expertise in urban risk reduction and adaptation planning are still scarce and fragmented. Key principles of traditional planning become questionable in times of increasing disasters and climate change. For example, bundled infrastructure (roads, telecommunication, water supply, etc.) that uses the same space or development corridors may reduce urban resilience, which relies upon redundancy and replaceable elements. Old solutions are not always enough to solve new problems. As Jürgen Nimptsch, Mayor of the City of Bonn, Germany and Vice Chair of the World Mayors Council on Climate Change (WMCCC) states:

> We city representatives need scientific expertise; we need consultancy; we need fresh thinking after spending years now trying to solve the problem.[1]

Sustainable urban development today needs to be based on planning practice that incorporates not only climate change mitigation, but also adaptation. Well-established 'sustainable city features' are, however, seldom defined in relation to risk reduction and adaptation. One of the main reasons is the complexity of adapting the urban fabric, which rules out the possibility of a one-size-fits-all solution. The fact that the suitability of any strategy and measure taken to foster risk reduction and adaptation has to be clarified in each individual case is, however, too often used as an excuse for missing integrative approaches in current planning practice.

Urban risk reduction and adaptation planning, as presented in this volume, is much more than identifying and acting on specific disaster and climate change impacts. It looks to the performance of each city's complex, dynamic and interconnected systems with their physical, environmental, socio-cultural, economic and political/institutional features that shape urban risk, and it includes the development of institutional and individual capacities to withstand or deal also with unexpected changes. Disaster resilience is thus not a static condition and risk reduction and adaptation planning is not a one-off action to eliminate urban risk, but an ongoing process which can be assessed by:

- The extent to which city authorities and other urban actors are success-ful in (improving capacities for) reducing current and future hazard exposure, vulnerability, and the shortages in response and recovery mechanisms and structures – with particular attention to how this serves at-risk citizens; and
- The degree to which they succeed in mainstreaming risk reduction and adaptation at institutional and interinstitutional levels to (a) back up measures taken at the programme level, (b) foster more distributed urban risk governance and (c) promote improved science–policy integration.

Such a risk reduction and adaptation process 'translates' into urban actors being able to reduce and adapt to evolving and changing risk in a flexible, dynamic and effective manner, enhancing the conditions and quality of city dwellers' lives. This includes the improvement of risk-reducing approaches to maintain systems functions and possibly also the transformation of sys-tems for achieving long-term sustainability.

Comprehensive risk reduction and adaptation planning incorporates the notion of 'transformation' and 'sustainability',[2] ensuring that related action taking contributes to more sustainable development pathways and enhanced human security. It provides capacities to transform fundamental attributes of urban systems (such as urban governance regimes, knowledge produc-tion systems and urban lifestyles) to shape future cities. In practice, it is likely that a combination of making relatively small changes to existing approaches and scaling-up efforts are required, but alongside transformative initiatives.

An important aspect in this context is the creation of a more distributed urban risk governance system, where the efforts of people and urban institutions can (better) support and complement each other. Whilst city authorities increas-ingly follow a planning process where the local knowledge and the climate impacts already observed are given some importance, this is not sufficient to develop long-term measures and strategies for risk reduction and adaptation planning. Urban risk governance is the combined domain wherein knowledge about disasters, climate change and urban planning, and their management, is coordinated, mediated and altered though joint governance practices. A higher diffraction or distribution of power, which also includes citizens themselves, can trigger more effective learning in a context of rapid change, and provide a better platform for 'experimenting' where conditions are uncertain and situations are in a constant state of flux. It also allows 'home grown' approaches and methods to evolve, and increases understanding and addressing of local causes of harm through participatory, community-based efforts formulated within the larger policy context.

City authorities and urban planning provide a powerful and vital platform for supporting risk reduction and adaptation. There is ample scope for potentially beneficial planning interventions that can foster more disaster-resilient cities and ultimately lead to sustainable urban risk governance and transformation. This volume compiles important knowledge required to tap into this potential: *Chapter 2* provides an overview of key concepts central to the understanding of urban risk reduction and adaptation. It integrates perspectives from disaster and climate change management approaches with a specific urban focus. Whilst there are no universally accepted definitions, it is crucial for urban actors to 'construct' a framework that can guide the comprehensive management of increasing urban disasters and changing risk patterns in practice. The succeeding *Chapter 3* describes the bidirectional relationship between disasters and cities, showing how the urban fabric with its characteristic urban features relates to both climate- and non-climate-related risk. In-depth knowledge on the city–disasters nexus is an indispensable precondition for urban actors, allowing them to modify current planning practice so as to act upon increasing risk and, where possible, address the particular root causes of urban risk. *Chapters 4* and *5* give an overview of current risk reduction and adaptation approaches taken by city authorities and city dwellers from diverse contexts in low-, middle- and high-income nations. They provide an array of exemplary measures and strategies from which to learn. *Chapter 5* further shows the importance of analysing local coping systems and taking citizens' own efforts into account when providing policies and projects to enhance disaster resilience and transformation. The subsequent *Chapter 6* summarizes current key challenges for risk reduction and adaptation and, on this basis, provides planning principles for addressing these. It is a kind of mental map that can help urban actors to guide the process of (mainstreaming) risk reduction and adaptation planning.

Put together, the different chapters of this volume provide effective insights into key theoretical concepts, supported by practical examples and suggestions for applying integrative approaches to risk reduction and adaptation in urban contexts. They advance current thinking to highlight ways in which resilient cities can be built and sustainable urban transformation achieved through extending current positive actions and addressing gaps and shortfalls in theory and practice.

Regardless of cities' context-specific baseline conditions, and independent of the specific strategies or measures taken, the following issues can be highlighted:

- Urban planning has the potential to act as an entry point and catalyst for the promotion of integral, participative and locally based risk reduction and adaptation.
- There is a range of valuable lessons to be learnt from (other) cities, including cities in high-, middle- and low-income countries. Knowledge and sustainable solutions can flow in both directions between and within the northern and southern hemispheres. In the context of risk reduction and adaptation planning, knowledge transfer is likely to require greater emphasis on processes and institutions as opposed to a focus on the transfer of 'hardware' (when compared with other fields such as climate change mitigation).
- The establishment of learning mechanisms and iterative decision-making is essential to enhance action, including creative and effective forms of knowledge transfer and exchange.
- Risk reduction and adaptation have to be mainstreamed into the context of urban development, response and recovery and thus have to be applied before, during and after the occurrence of (potential) hazard impacts.
- Risk reduction and adaptation measures need to consider different time periods and scales, i.e. near- and long-term as well as the building, neighbourhood, conurbation, catchment, regional, national and global scales, hereby also taking account of the close linkages between urban and rural environments. Urban–rural linkages are manifold and a reminder that risk reduction and adaptation should not be seen as isolated urban or rural agendas.
- All urban stakeholders, not least city dwellers, have the right to information on risk reduction and adaptation that is tailored to them.
- Special consideration needs to be given to vulnerable groups for whose security a particular urban context and associated disasters and climate change represent specific threats (such as the poor, children or women).

- It is not necessarily the effectiveness of citizens' single coping strategies, but the flexibility and inclusiveness of local coping systems (i.e. the combined set of strategies used) that is essential for building resilience and moving towards urban sustainable transformation. Support of urban residents to (negotiate their needs and rights in order to) increase the flexibility and inclusiveness of local coping systems, and make them more viable in today's context, is thus vital.

In sum, making cities safe from disasters and climate change impacts and enhancing citizens' security is everybody's business and part of the larger sustainability challenge. Comprehensive urban risk reduction and adaptation, as described and promoted in this book, demands sustainability science on a 'grand' scale, with new forms of collaborations and strong ambitions on the part of participants, including practitioners, researchers and citizens. Whether or not the multitude of small-scale private decisions contribute to increasing cities' resilience, and put us on a course towards sustainability, strongly depends on what city authorities do, encourage, support and prevent, and how well their work is backed up by regulatory and policy frameworks and research at local, national, regional and international levels. Ultimately, it is not a debate between the merits of inherent (local) and external (institutional) systems, but a question of finding the most appropriate approach to each situation.

I now have to close this book. Every page has been a journey. The journey is not finished. Only the book is.

Notes

Chapter 1: Setting the scene

1 For simplicity, the term climate change in this book refers to both climatic extremes and climatic variability. Climate change is defined as 'A change in the state of the climate that can be identified (e.g. by using statistical tests) by changes in the mean and/or the variability of its properties and that persists for an extended period, typically decades or longer. Climate change may be due to natural internal processes or external forcings, or to persistent anthropogenic changes in the composition of the atmosphere or in land use' (IPCC 2012b: 557). Note that the term climatic variability refers in this book also to changing weather patterns (see endnote 3).

2 This is despite the lack of global data as regards urban disasters, owing to the lack of emphasis on the urban context in development and disaster risk research and policy (Pelling 2007).

3 For simplicity, the terms climate conditions, climatic extremes and climatic variability refer in this book also to weather conditions, extreme weather events and changing weather patterns. Weather describes the condition of the atmosphere over a short period of time, e.g. from day to day or week to week, while climate describes average conditions over a longer period of time. In accordance, '[t]he distinction between extreme weather events and extreme climate events is not precise, but is related to their specific time scales. An extreme weather event is typically associated with changing weather patterns, that is, within time frames of less than a day to a few weeks. An extreme climate event happens on longer time scales. It can be the accumulation of several (extreme or non-extreme) weather events (e.g. the accumulation of moderately below average rainy days over a season leading to substantially below average cumulated rainfall and drought conditions)' (IPCC 2012a: 117).

4 http://www.emdat.be

5 The classification of nations into 'low-', 'middle-' and 'high-income economies' is supported by the World Bank, which sets a new classification every year on 1 July based on the previous year's gross national incomes (GNI) per capita. In line with their definition, a low-income economy has a GNI per capita below $1,025, a (lower to upper) middle-income economy between $1,026 and $12,475, and a high-income economy $12,476 or above (see http://data.worldbank.org/about/country-classifications). Low- and middle-income nations are sometimes

referred to as 'developing countries' or the 'global south', although definitions, and thereby the countries in each category, differ slightly. In a similar manner, high-income nations are sometimes referred to as 'developed countries' or the 'global north'.

6 Note that in this book, the terms 'urban planning' and 'city planning' are used as synonyms. Physical or spatial planning is considered to be part of urban planning.

7 See also the *Rio 2012 issues briefs no. 5* on Sustainable Cities (UNCSD 2012).

8 http://www.uncsd2012.org/7issues.html

9 http://www.ec.europa.eu/environment/eia

10 The term 'physical' refers in this book mainly to the shape and the structural or constructive aspects of the urban fabric. Physical changes include engineering or constructive measures as well as other large and small-scale changes of the urban fabric. Such changes are often referred to as 'hard measures' (as opposed to 'soft measures').

11 The term root causes is here defined as an interrelated set of structural factors and processes within a society, which often have arisen in another time (i.e. in the past) or another place (i.e. in a distant centre of economic or political power), and are typically so entrenched in today's society that they become 'invisible' and thus hard to detect. Root causes of poverty and risk often overlap to a great extent and include issues such as social, economic and political marginalization, inequality and exclusion, which become visible in unsafe conditions (e.g. of the urban fabric). In this book, root causes of risk refer also to intrinsic and entangled dynamic pressures (see Pressure and Release Model by Wisner *et al.* 2004).

12 The deliberately broad and inclusive scope is an important strength of this book; however, it also results in some limitations. The diversity of past events, processes and responses described had to be generalized so as not to detract from the key arguments and flow.

13 The term 'urban actors' is used as an umbrella term, which includes urban institutions as well as other stakeholders (including urban dwellers) that contribute to the formal and informal development planning of urban communities. See also the following endnote.

14 'Urban institutions' or 'urban organizations' refer in this book mainly to city authorities and other governmental or non-governmental organizations that work in the field of urban planning. This includes national governments, international, regional and subregional organizations and the private sector. The terms 'organization' and 'institution' are used as synonyms.

15 In this book, the terms 'city planning' and 'urban planning' are used as synonyms.

Chapter 2: Sorting out the conceptual 'jungle' associated with urban risk reduction and adaptation

1 The conceptual framework presented in this chapter is based on many years of teaching the subject and the revision of numerous publications (e.g. Adger *et al.* 2004, 2005; ADPC 2006; Ballard *et al.* 2008; Bendimerad *et al.* 1999; Benson *et al.* 2007; Bicknell *et al.* 2009; Cannon and Müller-Mahn 2010; CARE 2009; Comfort *et al.* 2010; Coppola 2011; da Silva *et al.* 2012; FAO/ILO 2009; Holden 2004; IISD 2007; IOI 2012; IPCC 2007a, 2012a; IRP 2010a, 2010b; ISO 2009;

Kates *et al.* 2012; Kim 1992; La Trobe and Davis 2005; Matthew *et al.* 2010; Mercer 2010; Mitchell 2003; Molin Valdes *et al.* 2013; Morss *et al.* 2011; O'Brien 2011; O'Brien and Leichenko 2008; Oxfam 2002; Pelling and Manuel-Navarette 2011; Plan International 2012; SDC 2009; Stephenson *et al.* 2010; Tearfund 2011, 2009; Tehler 2012, 2013; Turpeinen *et al.* 2008; Twigg 2004; UKCIP 2007; UN-IATF/DR 2006; UNAIDS 2008; UNISDR 2005, 2008a, 2008b, 2009, 2010a; Wilbanks and Kates 2010; Wisner *et al.* 2004, 2011). Only when statements in the following text are related to a particular aspect included in one of these publications will a reference be included in the text.

2 'A livelihood comprises the capabilities, assets (including both material and social resources) and activities required for a means of living' (DFID 1999: 1; Scoones 1998: 5; see Chambers and Conway 1991).

3 In the case of Copenhagen City, there is, for instance, no indication that poorer citizens are more vulnerable to floods (which seems to be contrary to the case of heat) (Leonardsen 2012).

4 There are certainly some differences between the variables presented in Figure 2.4, but the similarities are prominent.

5 Existing indices are, for instance: (a) The Disaster Risk Index (DRI) developed by UNDP (UNDP 2004); (b) the WorldRiskIndex (WRI) of Bündnis Entwicklung Hilft (Alliance Development Works) (http://www.worldriskreport.com); (c) the Climate Change Vulnerability Index (CCVI) by the global risks advisory firm Maplecroft (maplecroft.com/about/news/ccvi.html); (d) the Environmental Sustainability Index (ESI), developed in cooperation with Yale and Columbia Universities (e.g. http://www.yale.edu/esi); (e) the Environmental Vulnerability Index (EVI) of the South Pacific Applied Geosciences Commission (SOPAC) (http://www.vulnerabilityindex.net) (see Yohe *et al.* 2006); and (f) the Vulnerability–Resilience Indicator Prototype (VRIP) developed by Brenkert and Malone (Brenkert and Malone 2005).

6 The increased use of the term resilience is also a shift in communication towards more positive connotations, by turning away from the terms risk and risk reduction to discussions on how to foster resilience, as stated by Margareta Wahlström, Special Representative of the UN Secretary General for Disaster Risk Reduction (Wahlström 2012).

7 Adequate risk analyses are a necessary prerequisite to achieve this goal (see Box 2.2 and Section 2.2).

8 See also Chapter 1 endnote 11.

9 Note that response and recovery is often implemented *ad hoc* and is thus not always prepared.

10 The term coping strategy is, for instance, used in Béné *et al.* 2012; DFID 2004; Douglas *et al.* 2008; Gaillard *et al.* 2010; Soltesova *et al.* 2012; Twigg 2004; Wisner *et al.* 2004. The term coping mechanism is used in IPCC 2012a; individual practice for risk reduction in Shaw *et al.* 2008; private adaptation in IPCC 2007b; autonomous adaptation in Hamza *et al.* 2012 and IPCC 2007b; adaptive response in Dodman and Mitlin 2011; adaptive behaviour and adaptive practice in Banks *et al.* 2011; CARE 2001; Pelling *et al.* 2008; and the term indigenous practice in Shaw *et al.* 2008.

11 Note that some criticism has been directed towards risk matrices as a tool, indicating that they should be used with caution and only with careful explanation of which assumptions are made during the analysis. Cox (2008) presents a number of mathematical limitations to risk matrices, such as their possibility of assigning identical ratings to quantitatively very different risks (poor resolution),

assigning higher qualitative ratings to quantitatively smaller risks (errors) and their need for subjective interpretation of both input and output, resulting in the possibility that different users may obtain opposite ratings of the same quantitative risks (ambiguous inputs and outputs).

12 The aspects listed here are part of the so-called STAPLEEE method, a cost–benefit analysis tool, which considers social, technical, administrative, political, legal, environmental and economic issues (see FEMA 2003: 40, 92).

13 Only the mainstreaming concept has not yet been discussed because it will be explained in Section 2.3 and in more detail in Section 2.4.

14 Due to the historical development of different terms, disaster risk management and disaster management are partially used as synonyms. The same applies to the terms disaster risk management and disaster risk reduction. See Box 2.5. Originally, disaster management only included response and recovery.

15 When dealing with external assistance in so-called developing countries, these phases can also be called the 'traditional fields of international cooperation'.

16 Social protection is initiatives that aim to protect the livelihoods of socially excluded and marginalized people and to enhance their social status and rights (Devereux and Sabates-Wheeler 2004). Such initiatives can also be taken in response to levels of vulnerability, risk and deprivation that are deemed socially unacceptable within a given polity or society (Conway et al. 2000). See also Chapter 4 endnote 37.

17 Security is here understood as 'protecting the vital core of human life in ways that enhance human freedom and fulfillment' (Redclift et al. 2011: 6; see Bohle 2007: 14). In the context of increasing disasters and climate change, human security can be considered as a state or condition where 'individuals and communities have the options necessary to end, mitigate or adapt to threats to their human, environmental, and social rights; have the capacity and freedom to exercise these options; and actively participate in attaining these options' (Lonergan et al. 1999: 18; see Matthew et al. 2010: 18 and Redclift et al. 2011: 93).

18 The conceptual mainstreaming framework presented here was originally developed in the context of a research project carried out during 2003–2007 (Wamsler 2007c) and has since then been constantly refined.

19 This relates to work in the context of response, recovery and development.

20 Note that in the mainstreaming context, the terms 'measures for risk reduction' and 'measures for climate change adaptation' are somewhat misleading ways to describe how risk can be reduced because risk reduction and adaptation are generally not the main or single objective of related measures. The wording is, however, kept in this book for the sake of simplicity.

21 Urban planning is often categorized as one of many sectors, although this does not recognize its coordinating function, which makes it rather an area that relates to different sectors (Greiving and Fleischhauer 2012).

Chapter 3: The city–disasters nexus: a two-way relationship

1 Personal communication sent by email on 17 February 2012 to a group of urban experts.

2 The following sections on city features are based on many years of teaching and working on the subject, including the revision of numerous publications. Both

recent and decades-old literature on urbanization, city ecology and urban climate/thermal comfort were reviewed to analyse the links between urban form, climate and its relation to risk (e.g. Adam 1988; Alexander 2010; Baehring 2011; Baker 2012; Benton-Short and Short 2008; Bicknell *et al.* 2009; Bosher 2008; Brenner and Keil 2006; Bulkeley 2013; Bulkeley and Betsill 2003; da Silva *et al.* 2012; Dodman 2009; EEA 2012; Emmanuel 2005; Givoni 1998; Hall and Pfiffer 2000; IFRC 2010; Kay 1982; Koch-Nielsen 2002; Konya and Swanepoel 1980; LeGates and Stout 2000; McGranahan *et al.* 2007; Mitchell 2003; Mumford 1968; O'Brien and Leichenko 2008; Olgay 1963; Pelling 2003; Pelling and Wisner 2009; Roaf *et al.* 2005; Roberts *et al.* 2009; Salmon 1999; Sanderson 2000; Satterthwaite 2007, 2008; Shaw and Sharma 2011; Simmel 1960; Tacoli 2012; UNHABITAT 2007, 2010a, 2011; Weber 1966; Wirth 1938; Wisner 2003; Worldwatch Institute 2007; Yu 2006). Only when statements in the following text are related to a particular aspect included in one of these publications will the reference also be included in the text.

3 'In slums, risky sexual behaviour among women and girls, or trading sex for food or cash, is a widespread strategy to make ends meet. This makes females vulnerable to HIV/AIDS and other sexually transmitted diseases' (UNHABITAT 2010b: 18).

4 Leaving home before dawn or returning after nightfall, walking alone between bus stations or waiting for transport to arrive, they are an easy target for violent crimes (Tacoli 2012).

5 See Chapter 3 endnote 2.

6 Urban poverty is a complex phenomenon, which is sometimes referred to as 'new' poverty, with the 'new' meaning that it is related not necessarily to a lack of food and basic services but also (or rather) to aspects such as violent, stigma, family breakdown, drug issues and environmental degradation (*The Economist* 2010).

7 If they exist, social cohesion and leadership structures, with local leaders who have a deep commitment to the community and knowledge about its intrinsic social relations, can support the adequate identification of existing risk, needs and vulnerable groups, as well as improve access to and adequate distribution of assistance. The depth of commitment to the community and knowledge of social relations among informal leaders can also lead to important roles for women's, youth and faith-based groups in dealing with extreme situations (Pelling and Wisner 2009).

8 The analysis of past disaster impacts on the urban fabric is, in practice, also an important exercise to identify the influence of the urban fabric on disaster occurence (i.e. the reverse interrelation discussed in Section 3.2) because disaster impacts can bring former (pre-disaster) aspects of risk and vulnerability to light which are difficult to detect during 'normal' times.

9 http://www.unisdr.org/campaign/resilientcities

10 http://www.emdat.be

11 The induced cooling can immediately lower the global temperature before it returns to the warming trend and, therefore, climate change can become delayed by several decades (Tjiputra and Otterå 2011).

12 Although cities that concentrate wealthy people with high consumption lifestyles generally consume more energy than the poor in low- and middle-income nations, some cities in high-income countries, such as Stockholm (Sweden) and Barcelona (Spain), now produce fewer carbon emissions than some cities in low-income countries, for instance in Africa (UNHABITAT 2011, 2008).

13 Broadcast on 3 April 2008: http://www.bbc.co.uk/programmes/b009n3wf

Chapter 4: City authorities' approaches to urban risk reduction and adaptation

1   This chapter is based on many years of teaching and working on the subject, including the revision of numerous publications. The analysis has included both cross-country studies and country- and city-specific studies as well as further literature to compare and complement the findings (e.g. Alam and Rabbani 2007; BaltCICA 2012; Benson *et al.* 2007; Berke 1995; Bicknell *et al.* 2009; Brooks *et al.* 2009; CAP 2007; Carmin *et al.* 2012; CoR 2011; Davoudi *et al.* 2009; DFID n.d; DWR California 2008; Easterling *et al.* 2004; EEA 2012; European Commission 2009; European Union 2011; Fujikura and Kawanishi 2010; Gagnon-Lebrun and Agrawala 2006; Greater London Authority 2010; Goethert *et al.* 1992; Greiving and Fleischhauer 2012; Hagelsteen and Becker 2013; Hamdi 2010; Hamdi and Goethert 1997; Hardoy *et al.* 2011; IFRC 2004, 2010; IPCC 2000; IRP 2010a; Jha *et al.* 2012; Jha 2006; Johnson and Wilson 2006; Jonsson *et al.* 2011; Jürgens 2011; Kazmierczak and Carter 2010; Kleinfield 2011; Krantz 2001; La Trobe and Davies 2005; Leonardsen 2012; Loftus *et al.* 2011; Lourenço *et al.* 2009; Marulanda *et al.* 2010; Massey and Bergsma 2008; Matthies *et al.* 2008; Mechler and Bayer 2006; Meister *et al.* 2009; Mickwitz *et al.* 2009; Molin Valdes *et al.* 2013; MSB 2012a; NYCDOB 2011; Panda 2011; Pelling 2007; Rauhala and Schultz 2009; RedDesastres n.d.; Ribeiro *et al.* 2009; Roberts *et al.* 2012; Rode *et al.* 2012; Roggema 2009; Satterthwaite 2011a, 2011b; Satterthwaite *et al.* 2007a, 2007b; Schuster 2008; Shah and Ranghieri 2012; SKL 2011; Stein 2010; Swart *et al.* 2010; *The Economist* 2010; UNFCCC 2007; UNHABITAT 2011; UNISDR 2004, 2010a, 2010b, 2011a, 2012b, 2012c; USGS n.d.; and the adaptation portals http://www.ukcip.org.uk and http://www.stadtklimalotse. net). Only when statements in the following text are related to a particular aspect included in one of these publications will the reference be included in the text.

2   A brief overview of current measures is further provided in Section 6.1.

3   http://www.pdc.org/iweb/preparedness_information.jsp?subg=2

4   http://clerk.ci.seattle.wa.us/public/toc/t18.htm

5   http://www.inteligenciascolectivas.org/que-verde-era-mi-valla-publicitaria

6   http://www.getprepared.gc.ca/index-eng.aspx

7   http://www.lund.se and http://radishtradgard.wordpress.com/about

8   Web seminar on 'Urban agriculture and food security' from ICLEI Resilient Cities Team: http://resilient-cities.iclei.org/resilient-cities-hub-site/webinar-series/urban-agriculture-and-food-security

9   The term 'moral hazard' is generally used to describe the situation in which someone who is insured against disaster risk will ignore such risk, or even engage in risky behaviour on purpose, knowing that any costs incurred will be compensated by the insurer.

10   http://www.fema.gov/rebuild/recover/after.shtm

11   A brief overview of current measures is further included in Section 6.1.

12   The use of risk assessments is discussed in detail in the following section.

13   A brief overview of current mainstreaming strategies is further provided in Section 6.1.

14   As an example, one of the gaps in the federal government's adaptation programming in Canada was the intention to address climate change adaption separately from institutional aspects. The government had not clarified the management

and the strategies of the adaptation efforts, and which department would assume related responsibilities (Brooks *et al.* 2009).

15 http://www.salford.gov.uk/floodrisk-planguidance.htm

16 http://www.ncga.state.nc.us/gascripts/BillLookUp/BillLookUp.pl?Session= 2011&BillID=H819&submitButton=Go

17 http://www.capetown.gov.za/en/DRM/Pages/default.aspx and http://www. capetown.gov.za/en/DRM/Pages/Legislation.aspx

18 This has, for instance, been the case for County Administrative Boards in Sweden (Carlsson-Kanyama 2012).

19 *The territorial state and perspectives of the European Union* includes guidelines for improving climate change adaptation (European Union 2011). Whilst not many of these have been considered yet, the promoted integrated risk assessment is among the more widely used tools (Greiving and Fleischhauer 2012).

20 http://eur-lex.europa.eu/LexUriServ/LexUriServ.do?uri=OJ:L:2007:288:0027: 0034:EN:PDF

21 http://www.foi.se/sv/Kunder–Partners/Projekt/Climatools/Climatools/

22 http://www.ukcip.org.uk/lclip

23 http://www.ukcip.org.uk

24 http://www.stadtklimalotse.net

25 http://www.klimatanpassning.se

26 http://www.klimatilpasning.dk

27 http://safehospitals.info/index.php?option=com_content&task=view&id= 30&Itemid=1

28 An example is the Humboldt State University in northern California, which is sending students to the Dominican Republic where they are working to improve wind power systems, rain catchment systems and school buildings with locally available materials and knowledge to better withstand the challenges of the Dominican environment and its climatic conditions. See http://www.treehugger. com/green-architecture/dominican-and-american-students-build-school-waste-products-and-concrete.html

29 http://www.circle-era.eu

30 http://www.lucsus.lu.se and http://utmaninghallbartlund.se

31 http://www.isumalmo.se

32 http://idea.unalmzl.edu.co

33 Note that a brief overview of current mainstreaming strategies is also provided in Section 6.1.

34 Another good example is Copenhagen.

35 The description of the case of Manizales is based on the author's research undertaken in Colombia and was complemented with information from Cardona 2005, Hardoy *et al.* 2011 and Marulanda 2000.

36 Note that the term 'direct' means that citizens' risk is here directly reduced through institutional measures (as opposed to indirect risk reduction, i.e. the capacity development of at-risk citizens to be able to reduce and adapt to risk themselves; see Section 2.2).

37 On the basis of the sustainable livelihoods approach, also new concepts such as the adaptive social protection (ASP) have emerged, with the idea of combining key elements of social protection, risk reduction and adaptation approaches as a means to promote climate-resilient livelihoods in policy and practice in developing countries (IDS 2012; Practical Action 2010). See also Chapter 2 endnotes 2 and 16.

38 http://www.texxi.com

39 http://www.cicloviasrecreativas.org/en
40 http://www.nyc.gov/html/ceo/html/programs/opportunity_nyc.shtml
41 The list shows that in contrast to climate change mitigation, knowledge transfer for risk reduction and adaptation requires greater emphasis on processes and institutions (as opposed to a focus on the transfer of 'hardware'). This is not only true for south–north knowledge transfer.
42 http://www.citynet-ap.org/programmes/services/technical-cooperation-among-cities-of-developing-countries-tcdc
43 http://www.unisdr.org/campaign/resilientcities/cities/index4/#view_role_model
44 An example is the Sustainable Value Creation Project (*Hållbart värdeskapande*), initiated in 2009 by 14 of Sweden's largest institutional investors, which promotes a structured approach towards sustainability. In the concept of sustainability they include environment (climate change mitigation only, not adaptation). The project included a census among the 100 biggest companies in Sweden to map how far they had come towards sustainability. Regarding mitigation, the questions addressed, for example, whether the companies had any regulations regarding their impacts on the environment (and more specifically, regarding emissions, consumption of resources [water, energy, etc.], waste and wastewater management and biodiversity).
45 http://www.kk.dk/CityOfCyclists
46 http://www.jardins-familiaux.org
47 http://web.comhem.se/~u36802589
48 An example of a city that promotes urban agriculture as a means for adaptation is Lund in Sweden. See http://www.lund.se (*Brunnshög* and *Odla i Lund* projects).

Chapter 5: City dwellers' own ways to reduce and adapt to urban risk

1 Migration to places of better opportunities and the modification of diet to accommodate a changing food endowment form part of human history and development.
2 Note that some scholars differentiate between coping strategies (as being short-term local adjustments to deal with extreme weather events) and adaptive behaviour (as being longer-term or more fundamental changes people make to systematically reduce potential harm or take advantage of opportunities from changing climate events) (e.g. Cutter *et al.* 2008; Gallopín 2006; IPCC 2012a; Smit and Wandel 2006). These two terms are used here as synonyms because their differentiation is not clear-cut and is very context-specific. In fact, from a practical perspective, coping and adaptation overlap significantly, are closely connected, have both synergies and trade-offs, and thus related capacities might complement each other (see Béné *et al.* 2012; Wamsler 2007a).
3 Also called anticipatory or reactive coping strategies (IPCC 2007a).
4 Statement by Lennart Cegrell. See http://hd.se/helsingborg/2011/11/28/vattnet-bara-forsade-in-i-vart-hus
5 The term 'physical' refers in this book mainly to the shape and the structural or constructive aspects of the urban fabric. Physical changes include engineering or constructive measures as well as other large and small-scale changes of the

urban fabric. Such changes are often referred to as 'hard measures' (as opposed to 'soft measures').

6 This categorization has been established by 'cultural theory' (Thompson *et al.* 1990). The theory, developed originally by Mary Douglas and expanded by Thompson and his colleagues, argues that there are only four ways of organizing, perceiving and justifying social relations: hierarchy, individualism, egalitarianism and fatalism. These four 'ways of life' conflict in every conceivable social domain because each will define both the problem and the solution in a specific way that contradicts the others. There are, in consequence, four 'voices'.

7 More generally, 'the hierarchist voice is pro-control. It talks of "global stewardship", is quick to point out that what is rational for the parts may be disastrous for the whole, and insists that global problems such as climate change demand global and expertly planned solutions. The individualist voice is pro-market. It calls for deregulation, for the freedom to innovate and take risks, and for the internalization of environmental costs so as to "get the prices right." The egalitarian voice is strident and critical. It scorns the idea of "trickle down," argues for zero-growth and calls for major shifts in our behaviour so as to bring our profligate consumption down within the limits set by Mother Nature. Fatalist actors see no possibility of effecting change for the better, and tend to have no voice. Even so, they have their vital part to play because we need to hear their counsel to not waste time and money over things about which we can do nothing' (Thompson 2011: 12).

8 Note that the differentiation between the different aspects proposed to analyse coping measures in a systematic way is not always clear-cut and there are many overlaps.

9 This chapter is based on many years of teaching and working on the subject and the revision of numerous publications. The analysis has included both cross-case and single-case studies, as well as further literature to compare and complement the findings (e.g. Adger 1996; Adger *et al.* 2004, 2005; Audefroy 2011; Australian Red Cross 2010; Ayers and Forsyth 2009; BaltCICA 2012; Banks *et al.* 2011; Ben Cheikh and Bouchair 2008; Cadot *et al.* 2007; Carmin *et al.* 2012; CWSPA 2005; DFID n.d.; Dodman and Mitlin 2011; Douglas *et al.* 2008; EEA 2012; Eriksen *et al.* 2011; Esdahl 2011; European Commission 2008; Few 2003; Guarnizo 1992; Hamza *et al.* 2012; Hasling 2012; IFRC 2010, 2004; Jabeen *et al.* 2010; Jha 2006; Khan 2008; Larsen 2006; Leonardsen 2012; Lindblad 2012; McAdoo *et al.* 2009; Meister 2009; Morduch 1999; Morss *et al.* 2011; O'Brien and Leichenko 2008; Pelling 2011; Pizarro 2009: 38; Reale 2010; RedDesastres n.d.; Risbey *et al.* 1999; Rowswell and Clover 2012; SAARC 2008; Satterthwaite 2011b; Satterthwaite *et al.* 2007a, 2007b; Schinkel *et al.* 2011; Shaw *et al.* 2008; Simatele 2010; Singh 2011; Soltesova *et al.* 2012; Steffen *et al.* 2004; SVS 2011; Thompson 2011; Thompson and Wildavsky 1986; Thompson *et al.* 2007; Tudehope 2011; Twigg 2004; UN Water 2007; UNISDR 2012b, 2012c; USGS n.d.; Wisner 2003; and Wisner *et al.* 2004). Only when statements in the following text are related to a particular aspect included in one of these publications will the reference be included in the text.

10 The quotations included at the beginning of this and the following sections, which do not include references, are fictive, i.e. they have been adapted for this book, but are all based on real situations and interviews conducted by the author.

11 http://www.inteligenciascolectivas.org/muros-vegetales-de-contencion

12 http://www.inteligenciascolectivas.org/desagues-en-voladizoduchas-urbanas

13 An example from the Dominican Republic: http://www.inteligenciascolectivas. org/vivienda-elevada

14 See e.g. Olukoya 2011 and http://www.treehugger.com/green-architecture/ nigeria-plastic-bottle-house.html. This technique is commercially supported.

15 http://www.sunsetent.com/Categories/quakesecure.html

16 http://www.sswm.info/category/implementation-tools/wastewater-treatment/ hardware/greywater/soak-pit

17 http://www.infojardin.com/foro/showthread.php?p=2376470 and http://www. infojardin.com/foro/showthread.php?t=70758

18 http://www.hippo-the-watersaver.co.uk

19 http://www.viresco.com/ppc/erosioncontrol.html?gclid=COq1zu756LACFY-WEDgodGjvc1w

20 Another coping strategy of poor urban dwellers that requires little prior investment and is adopted when other income sources are not available is to look through garbage for recyclable materials that can then be sold. Both recycling and urban agriculture are resources that can generally be accessed without money (i.e. outside the urban monetary market), as opposed to most other things in the city (such as water, food, building materials, etc.) (Pelling 2003). See also Section 3.1.

21 E.g. http://www.floodgate.ltd.uk

22 E.g. http://www.wwtsasia.com/sandbags.htm

23 In general, insurance cover is heavily concentrated in urban areas. While insurance cover in high-income countries is widespread, in low and lower middle-income countries it is principally the large-scale commercial sector and elite and middle classes who are able to afford insurance premium payments (Simon 2013).

24 The citizens' summit was carried out in Kalundborg, Denmark. See http://www.baltcica.org

25 Note that coping strategies in general, and also the related informality in low- and middle-income nations, are not sufficiently researched but are relatively well described in scientific papers when compared with the knowledge on the coping strategies of people living in high-income nations, which has to be extracted mostly on the basis of non-scientific sources.

26 http://www.civil.se/om-oss

27 There are also institutionally sponsored 'climate change champion' campaigns or programmes to support citizens in their actions for climate change mitigation. See e.g. http://www.walescarbonfootprint.gov.uk/youngpeople/champion/?lang= en and http://www.britishcouncil.org/japan-science-climate-change-projects-climate-change-champion.htm. In contrast, institutionally sponsored 'climate *adaptation* champion' campaigns or programmes to highlight the adaptation actions of some citizens are rare, with few exceptions. See http://www.nccarf. edu.au/engagement/nccarf-climate-adaptation-champions

28 Effectiveness relates here to success in preventing deaths and injuries, and loss and damage to property, the environment and livelihoods in the context of current and future hazards (see also Adger *et al.* 2005: 81).

29 See http://www.iiasa.ac.at/web/home/research/researchProjects/FutureSoc/ FutureSoc.en.html and http://www.oeaw.ac.at/wic/index.php?id=95

30 http://www.inteligenciascolectivas.org

31 Shack/Slum Dwellers International (SDI) is a network of community-based organizations of the urban poor in 33 countries in Africa, Asia and Latin America. It was launched in 1996 when 'federations' of the urban poor in countries such as India and South Africa agreed that a global platform could help their local

initiatives develop alternatives to evictions while also impacting on the global agenda for urban development. In 1999, SDI became a formally registered entity. See http://www.sdinet.org

32 Fundación Salvadoreña de Desarrollo y Vivienda Mínima (Spanish for Salvadorean Foundation of Development and Housing) (see also Box 5.4): http://www.fundasal.org.sv

33 Increased levels of formal education can provide the skills and social capital that enable people to flexibly find appropriate and forward-looking solutions to climate-induced problems as they arise and react to anticipated risks that have not yet become manifest (see Section 5.5).

Chapter 6: Advancing sustainable urban risk governance and transformation

1 Statement by Eric Klinenberg, author of the book *Heat wave: a social autopsy of disaster in Chicago* (Klinenberg 2003).

2 http://www.revistanos.cl/2010/01/a-cinco-anos-de-falsa-alarma-crecen-temores-por-amenaza-de-tsunami

3 'Essential 1: Put in place organization and coordination to understand and reduce disaster risk, based on participation of citizen groups and civil society. Build local alliances. Ensure that all departments understand their role to disaster risk reduction and preparedness. Essential 2: Assign a budget for disaster risk reduction and provide incentives for homeowners, low-income families, communities, businesses and public sector to invest in reducing the risks they face. Essential 3: Maintain up-to-date data on hazards and vulnerabilities, prepare risk assessments and use these as the basis for urban development plans and decisions. Ensure that this information and the plans for your city's resilience are readily available to the public and fully discussed with them. Essential 4: Invest in and maintain critical infrastructure that reduces risk, such as flood drainage, adjusted where needed to cope with climate change. Essential 5: Assess the safety of all schools and health facilities and upgrade these as necessary. Essential 6: Apply and enforce realistic, risk compliant building regulations and land use planning principles. Identify safe land for low-income citizens and develop upgrading of informal settlements, wherever feasible. Essential 7: Ensure education programmes and training on disaster risk reduction are in place in schools and local communities. Essential 8: Protect ecosystems and natural buffers to mitigate floods, storm surges and other hazards to which your city may be vulnerable. Adapt to climate change by building on good risk reduction practices. Essential 9: Install early warning systems and emergency management capacities in your city and hold regular public preparedness drills. Essential 10: After any disaster, ensure that the needs of the survivors are placed at the centre of reconstruction with support for them and their community organizations to design and help implement responses, including rebuilding homes and livelihoods' (UNISDR 2012b: 25).

4 Success factor 1: Awareness-raising should be targeted at stakeholders ranging from citizens to national and European governments and cover various aspects of climate change adaptation. Success factor 2: Successful adaptation cuts across sectors and scales (including horizontal and vertical collaboration). Success factor 3: Adaptation solutions that deliver additional benefits, such

as green urban areas, increase the willingness to accept and implement them. Success factor 4: Working with nature, instead of working against natural processes. Success factor 5: Many climate change adaptation measures can be implemented at low-cost or contribute positively in other areas. A sufficient resource base in terms of financial, human and institutional resources needs, however, to be developed and secured (EEA 2012: 79).

5 Even if urban organizations decide not to include disaster property insurance in social housing credits, they could lobby for (other) governmental organizations or commercial insurance firms to at least cover schools, bridges and hospitals.

6 A grey and green infrastructure approach consists of, for instance, the combination of construction measures using engineering services *and* vegetated areas, thus contributing to the increase of ecosystem resilience and delivery of ecosystem services. Green infrastructure consists of natural or semi-natural areas in cities, which include parks, forest areas, playing fields and other open spaces, vegetated road-sides and tree-lined roads, private gardens, rivers, canals, lakes, wetlands, wades and seashores, green roofs and green walls.

7 http://www.ushahidi.com/about-us

8 http://plasticbagbanreport.com/category/fee-only

9 http://www.communities.gov.uk/publications/planningandbuilding/codeguide

10 See http://www.acccrn.org/about-acccrn/acccrn-methodology/shared-learning-dialogues

11 As an example, the Swedish insurance company Skandia has set up a foundation called 'Ideas for life' (*Idéer för livet*), which encourages social and physical activities, self-esteem and inclusion for children and youths. Whilst related actions focus on people who may end up in social exclusion or criminality, a similar concept could be used to support at-risk citizens to reduce and adapt to the risk they face. See http://www.ideerforlivet.se

12 See http://www.nccarf.edu.au/engagement/nccarf-climate-adaptation-champions

13 Examples are insurance schemes where people who have carried out constructive mitigation work pay lower premiums than those taking no measures to reduce risk, or insurance policies to which access is conditional on risk reduction. As regards the latter, in Fiji a structural engineer must certify that houses have certain cyclone-resistant features before owners can access disaster property insurance. In addition, city authorities and/or international donors can provide support regarding aspects of technical and administrative insurance. They can also offer community insurance policies or reinsurance in cooperation with national NGOs and national or international insurers (Wamsler 2007a).

14 Some of the most telling evidence on crowding out comes from South Africa (1993), when the government extended basic pension benefits to black South Africans, replacing informal means of coping with ageing and economic downturns (Morduch 1999).

15 http://www.peepoople.com

16 Based on experience in the health sector, micro-insurance needs to be complemented by non-financial preventive measures to become successful. In the housing sector, disaster property insurance could be linked to preventive construction programmes that involve training of community construction workers or the establishment of local advisory services. Ideally, social housing organizations or housing financing institutions would offer risk reduction measures to ensure that credits are paid back and no insurance claims become necessary (Wamsler 2007a).

17 There are many examples as to how physical interventions to improve the urban fabric have been used as an entry point to achieve more far-reaching improvements (by capitalizing on the linkages between physical, environmental, social and economic aspects). Rowswell and Clover (2012) provide for instance an example as to how retrofitting and upgrading measures in Cape Town, South Africa and Maputo, Mozambique have led to improved health, education, livelihoods and ultimately resilience. See also Section 6.3.1 (in relation to planning principle 4).

18 As stated by Simon (2013: 1), the 'definitions of human security integrate concerns with both the ability to meet short term needs and rights, and longer term capacities and fulfilment. Put differently, human security is about realising – and having the capacity to realise – one's potential as an individual and as part of a wider community. This requires the exercise of human agency within broadly enabling societal and environmental contexts for the control of vulnerability and promotion of sustainability (of which the resilience of appropriate elements is one dimension).' See also final paragraph in Section 2.3.

Chapter 7: Concluding remarks

1 The statement was made in reference to the issue of climate change adaptation during Nimptsch's presentation at the Resilient Cities conference on 13 May 2012. See http://resilient-cities.iclei.org

2 This includes the disaster resilience of appropriate elements.

Bibliography

ABC (Australian Broadcasting Company), 2012. Wet forecast scares post-flood Thailand, Television broadcast. *ABC News*, 9 April 2012.

Adam, K., 1988. *Stadtökologie in Stichworten [Urban ecology in a nutshell]*. Hirts Stichwörterbuch: Ökologie: Unterägeri.

Adger, N., 1996. *Approaches to vulnerability to climate change*. CSERGE Working Paper GEC 96-05. Norwich: Centre for Social and Economic Research on the Global Environment, University of East Anglia, and London: University College London.

Adger, N., *et al.*, 2004. *New indicators of vulnerability and adaptive capacity*. Norwich: Tyndall Centre for Climate Change Research.

Adger, N., Arnell, N., and Tompkins, E., 2005. Successful adaptation to climate change across scales. *Global Environmental Change*, 15, 77–86.

ADPC (Asian Disaster Preparedness Center), 2006. *Participant's workbook: community-based disaster risk management for local authorities (CBDRM)*. Bangkok: ADPC.

Alam, M., and Rabbani, M., 2007. Vulnerabilities and responses to climate change for Dhaka. *Environment and Urbanization*, 19 (1), 81–97.

Alexander, D., 2010. The L'Aquila earthquake of 6 April 2009 and Italian government policy on disaster response. *Journal of Natural Resources Policy Research*, 2 (4), 325–342.

Amnesty International, 2010. *Risking rape to reach a toilet: women's experiences in the slums of Nairobi, Kenya*. London: Amnesty International.

Arrieta, O., 2012. Nueva York se pinta de blanco para refrescarse [New York paints itself white to cool down]. *BBC Mundo*, 13 August 2012.

Audefroy, J., 2011. Haiti: post-earthquake lessons learned from traditional construction. *Environment and Urbanization*, 23 (2), 447–462.

Australian Red Cross, 2010. *Traditional knowledge and Red Cross disaster preparedness in the Pacific*. Australian Red Cross and Returned Australian Youth Ambassadors for Development (RAYAD).

Ayala, L., 2012. Caso tsunami: La Fiscalía pide arraigo y firma para los formalizados [The tsunami case: prosecution asks for outward voyage restrictions to those involved]. *Emol Chile*, 7 May 2012.

Ayers, J., and Forsyth, T., 2009. Community based adaptation to climate change. *Environment: Science and Policy for Sustainable Development*, 51 (4), 22–31.

Ayers, J., *et al.*, eds., 2012. *Community-based adaptation: scaling it up*. London: Earthscan.

Baehring, A., 2011. Facing the urban challenge for emergency shelters: analysis of methodological and conceptual approaches adopted by members of the Emergency Shelter Cluster in Port au Prince after the 2010 Haiti earthquake. Thesis (Masters). Lund and Copenhagen University.

Baker, J., 2012. *Climate change, disaster risk, and the urban poor: cities building resilience for a changing world*. Washington, DC: The World Bank.

Ballard, D., Alexander, S., and Goldthorpe, M., 2008. *Adaptive capacity benchmarking matrix (working towards a draft 'Organisational Change Tool')*. Winchester: ESPACE.

BaltCICA., 2012. *BaltCICA: climate change impacts, costs and adaptation in the Baltic Sea region – final report*. Espoo: BaltCICA.

Banks, N., Roy, M., and Hulme, D., 2011. Neglecting the urban poor in Bangladesh: research, policy and action in the context of climate change. *Environment and Urbanization*, 23 (2), 487–502.

BBC Mundo., 2012. Juzgan en Chile a ocho personas por no alertar del tsunami de 2010. *BBC Mundo*, 7 May 2012.

Becker, P., Abrahamsson, M., and Tehler, H., 2011. An emergent means to assurgent ends: community resilience for safety and sustainability. *In*: E. Hollnagel, E. Rigaud, and B. Besnard, eds. *Proceedings of the fourth resilience engineering symposium*. Sophia Antipolis: MINES ParisTech, 29–35.

Ben Cheikh, H., and Bouchair, A., 2008. Experimental studies of a passive cooling roof in hot arid areas. *Revue des Energies Renouvelables*, 11 (4), 515–522.

Bendimerad, F., *et al.*, 1999. The earthquakes and megacities initiative. *Science International*, 70, 5–7.

Béné, C., *et al.*, 2012. Resilience: new utopia or new tyranny? *Reflection about the potentials and limits of the concept of resilience in relation to vulnerability reduction programmes*. IDS Working Paper, Vol. 2012, No. 405, CSP Working Paper No. 006. London: Institute of Development Studies and Centre for Social Protection.

Benson, C., Twigg, J., and Rossetto, T., 2007. *Tools for mainstreaming disaster risk reduction: guidance notes for development organizations*. Geneva: ProVention Consortium.

Benton-Short, L., and Short, J., 2008. *Cities and nature*. Abingdon and New York: Routledge.

Berke, P., 1995. *Natural hazard reduction and sustainable development: a global assessment*. Center for Urban and Regional Studies, University of North Carolina at Chapel Hill. Working Paper: S95-02.

Bicknell, J., Dodman, D., and Satterthwaite, D., 2009. *Adapting cities to climate change: understanding and addressing the development challenges*. London: Earthscan.

Bohle, H., 2007. *Living with vulnerability: livelihoods and human security in risky environments*. Bonn: United Nations University Institute for Environment and Human Security, Intersections Publication Series, No. 6/2007.

Bosher, L., 2008. *Hazards and the built environment: attaining built-in resilience.* Abingdon and New York: Routledge.

Brenkert, A., and Malone, E., 2005. Modeling vulnerability and resilience to climate change: a case study of India and Indian states. *Climatic Change*, 72 (1–2), 57–102.

Brenner, N., and Keil, R., 2006. *The global cities reader.* New York: Psychology Press.

Brooks, M., *et al.*, 2009. *Prioritizing climate change risks and actions on adaptation: a review of selected institutions, tools and approaches: final report.* Ottawa: Policy Research Initiative.

Bulkeley, H., 2013. *Cities and climate change.* Routledge critical introductions to urbanism and the city. New York: Routledge.

Bulkeley, H., and Betsill, M., 2003. *Cities and climate change: urban sustainability and global environmental governance.* London and New York: Routledge.

Byrnes, M., 2012. The growing popularity of Bus Rapid Transit. *The Atlantic Cities: place matters. Focus: sustainability*, 1 June 2012.

Cadot, E., Rodwin, V., and Spira, A., 2007. In the heat of the summer: lessons from the heat waves in Paris. *Journal of Urban Health*, 84 (4), 466–468.

Cannon, T., and Müller-Mahn, D., 2010. Vulnerability, resilience and development discourses in context of climate change. *Natural Hazards*, 55 (3), 621–635.

CAP (Clean Air Partnership), 2007. *Cities preparing for climate change: a study of six urban regions.* Toronto: CAP.

Caparelli, M., Palazzo, L., and Kone, R., 2012. In Brazil, adolescents use UNICEF's new digital mapping technology to reduce disaster risks in the favelas. *UNICEF Newsline.* Rio de Janeiro: UNICEF Brazil, 19 March 2012.

Cardona, O., 2005. *La gestión del riesgo colectivo: un marco conceptual que encuentra sustento en una ciudad laboratorio [The management of collective risk: a conceptual mark which finds support in a laboratory city].* Manizales: Instituto de Estudios Ambientales, Universidad Nacional de Colombia.

CARE, 2001. Background paper prepared for the Experts Workshop on Participatory Monitoring and Evaluation for Community-based and Local Adaptation, 17–18 February 2011, London, UK.

CARE, 2009. *Climate vulnerability and capacity analysis handbook.* Geneva: CARE International.

Carlsson-Kanyama, A., 2012. *CLIMATOOLS.* Lecture for the course MVENI18 Klimatstrategiska metoder [Methods for climate change adaptation and mitigation] at Lund University. 21 May 2012, Lund, Sweden. See also http://www.foi.se/sv/Kunder–Partners/Projekt/Climatools/Climatools/

Carmin, J., Nadkarni, N., and Rhie, C., 2012. *Progress and challenges in urban climate adaptation planning: results of a global survey.* Cambridge, MA: MIT.

Carwright, A., Brundrit, G., and Fairhurst, L., 2008. *Sea-level rise adaptation and risk mitigation measures for the City of Cape Town.* Cape Town: LaquaR Consultants CC.

CBC News, 2007. Blowing in the wind: global moves against shopping bags. *CBC News*, 28 March 2007.

Chambers, R., and Conway, G., 1991. *Sustainable rural livelihoods: practical concepts for the 21st century*. IDS Discussion Paper 296. Brighton, UK: IDS.

Clark, W., and Dickson, N., 2003. Sustainability science: the emerging research program. *Proceedings of the National Academy of Sciences*, 100 (14), 8059–8061.

Comfort, L., Boin, A., and Demchak, C., eds., 2010. *Designing resilience: preparing for extreme events*. 1st ed. Pittsburgh, PA: University of Pittsburgh Press.

Conway, T., De Haan, A., and Norton, A., 2000. *Social protection: new directions of donor agencies*. London: Department for International Development.

Coppola, D., 2011. *Introduction to international disaster management*. 2nd ed. London: Elsevier.

CoR (Committee of the Regions), eds., 2011. *Adaptation to climate change: policy instruments for adaptation to climate change in big European cities and metropolitan areas*. Brussels: European Union, Committee of the Regions.

Cox, L., 2008. What's wrong with risk matrices? *Risk Analysis*, 28 (2), 497–512.

Cutter, S., *et al.*, 2008. A place-based model for understanding community resilience to natural disasters. *Global Environmental Change*, 18 (4), 598–606.

CWSPA (Church World Service Pakistan/Afghanistan), 2005. *Participatory research on indigenous coping mechanisms for disaster management in Mansehra and Battagram districts, North West Frontier Province (NWFP), Pakistan*. Church World Service – Pakistan/Afghanistan: World Vision International.

da Silva, J., Kernaghan, S., and Luque, A., 2012. A systems approach to meeting the challenges of urban climate change. *International Journal of Urban Sustainable Development*, 4 (2), 1–21.

Davoudi, S., Crawford, J., and Mehmood, A., eds., 2009. *Planning for climate change: strategies for mitigation and adaptation for spatial planners*. Abingdon and New York: Routledge.

De Cordier, B., 2011. The position and role of faith-based organisations in DRR, Box 10.2. *In*: D. Chester, A. Duncan, and H. Sangster, eds. *The Routledge handbook of hazards and disaster risk reduction*. Abingdon and New York: Routledge, 118.

Devereux, S., and Sabates-Wheeler, R., 2004. *Transformative social protection*. IDS Working Paper 232. Brighton: IDS.

DFID (UK Department for International Development), 1999. *Sustainable livelihoods guidance sheets*. London: DFID.

DFID (UK Department for International Development), 2004. *Disaster risk reduction: a development concern. A scoping study on links between disaster risk reduction, poverty and development*. London: DFID.

DFID (UK Department for International Development), n.d. *Improving knowledge transfer in urban development*. Output from the Urbanisation Knowledge and Research Project: Improving Research Knowledge Technical Transfer. London: DFID.

Dobson, S., 2011. Climate change mitigation in our cities: don't ignore the urban poor. *Daily Monitor*, 3 October 2011.

Dodman, D., 2009. Urban density and climate change. Analytical review of the interaction between urban growth trends and environmental changes paper 1. New York: United Nations Population Fund (UNFPA).

Dodman, D., and Mitlin, D., 2011. Challenges for community-based adaptation: discovering the potential for transformation. *Journal of International Development*, 25 (5), 640–659.

Dorell, O., 2012. Society not ready for heat waves coming with climate change. *USA Today*, 3 July 2012.

Douglas, I., *et al.*, 2008. Unjust waters: climate change, flooding and the urban poor in Africa. *Environment and Urbanization*, 20 (1), 187–205.

DWR California (Department of Water Resources), 2008. *Urban drought guidebook 2008: updated edition.* State of California Department of Water Resources Office of Water Use Efficiency and Transfers.

Easterling, W., Hurd, B., and Smith, J., 2004. Box 1: Adaptation in developing countries [p. 18], *Coping with global climate change: the role of adaptation in the United States.* Arlington, VA: Pew Center on Global Climate Change.

Economist, The, 2010. Brazil's Bolsa Familia: how to get children out of jobs and into school. *The Economist*, 29 July.

EEA (European Environment Agency), 2012. *Urban adaptation to climate change in Europe: challenges and opportunities for cities together with supportive national and European policies* (No. 2). Copenhagen: EEA.

Emmanuel, M., 2005. *An urban approach to climate-sensitive design: strategies for the tropics.* London: Taylor and Francis.

Eriksen, S., *et al.*, 2011. When not every response to climate change is a good one: identifying principles for sustainable adaptation. *Climate and Development*, 3 (1), 7–20.

Esdahl, S., 2011. Supporting societies' adaptive capacities to climate change: analysis and comparison of local and institutional capacities on Caye Caulker, Belize. Thesis (Masters). Lund and Copenhagen University.

European Commission, 2008. *Attitudes of European citizens towards the environment. Special Eurobarometer.* Brussels: European Commission.

European Commission, 2009. *Commission staff working document accompanying the White Paper – adapting to climate change: towards a European framework for action – impact assessment.* Brussels: European Commission.

European Union, 2011. *The territorial state and perspectives of the European Union.* Background document for the territorial agenda of the European Union 2020. Brussels: European Union.

FAO/ILO (Food and Agriculture Organization/International Labour Organization), 2009. *The livelihood assessment tool-kit (LAT): analysing and responding to the impact of disasters on the livelihoods of people.* Rome: FAO; Geneva: ILO.

FEMA (Federal Emergency Management Agency), 2003. *State and local mitigation planning how-to guide: developing the mitigation plan – identifying mitigation actions and implementation strategies.* Washington, DC: FEMA.

Few, R., 2003. Flooding, vulnerability and coping strategies: local responses to a global threat. *Progress in Development Studies*, 3 (1), 43–58.

Forsberg, B., 2012. *Samhällets och vårdens möjligheter att inom riskgrupper förebygga dödsfall under allt mer extrema värmeböljor [The potential of society and healthcare to prevent deaths in high-risk groups during increasingly extreme heatwaves]*, Project description, ongoing research project. Umeå: Department of Public Health and Clinical Medicine, Umeå University.

Fujikura, R., and Kawanishi, M., 2010. *Climate change adaptation and international development: making development cooperation more effective*. London and Washington, DC: Earthscan.

Gagnon-Lebrun, F., and Agrawala, S., 2006. Progress on adaptation to climate change in developed countries: an analysis of broad trends. *OECD Papers*, 6 (2), 8.

Gaillard, J.-C., 2010. Vulnerability, capacity, and resilience: perspectives for climate and development policy. *Journal of International Development*, 22 (2), 218–232.

Gallopín, G., 2006. Linkages between vulnerability, resilience, and adaptive capacity. *Global Environmental Change*, 16, 293–303.

Givoni, B., 1998. *Climate considerations in building and urban design*. New York: Van Nostrand Reinhold.

Goethert, R., et al., 1992. *La microplanificación. Un proceso de programación y desarrollo con base en la comunidad [Micro-planning. A community-based planning and development process]*. Washington, DC: IDE of the World Bank and FICONG.

Gonzalez, G., 2005. Urban sprawl, global warming and the limits of ecological modernisation. *Environmental Policies*, 14 (3), 344–362.

Greater London Authority, 2010. *The draft climate change adaptation strategy for London*. London: Greater London Authority.

Greiving, S., and Fleischhauer, M., 2012. National climate change adaptation strategies of European states from a spatial planning and development perspective. *European Planning Studies*, 20 (1), 27–48.

Guardian, The 2012. North Carolina tries to wish away sea-level rise. *The Guardian*, 1 June.

Guarnizo, C.C. 1992. Living with hazards: communities' adjustment mechanisms in developing countries. *In*: A. Kreimer and Y.M. Munansinghe, eds. *Environmental management and urban vulnerability*. Washington, DC: The World Bank, 93–106.

Hagelsteen, M., and Becker, P., 2013. Challenging disparities in capacity development for disaster risk reduction. *International Journal of Disaster Risk Reduction*, 3, 4–13.

Hall, P., and Pfeiffer, U., 2000. *Urban future 21: a global agenda for twenty-first century cities*. 1st ed. London and New York: Routledge.

Hamdi, N., 2010. *The placemaker's guide to building community: planning, design and placemaking in practics*. London and Washington, DC: Earthscan.

Hamdi, N., and Goethert, R., 1997. *Action planning for cities: a NY guide to community practices*. New York: John Wiley.

Hamza, M., Smith, D., and Vivekananda, J., 2012. *Difficult environments: bridging concepts and practice for low carbon climate resilient development*. Brighton, UK: IDS Learning Hub.

Hardoy, J., Pandiella, G., and Barrero, L., 2011. Local disaster risk reduction in Latin American urban areas. *Environment and Urbanization*, 23 (2), 401–413.

Hasling, A., 2012. *Is it economically worthwhile to adapt? The case of Copenhagen*. Presented at the Resilient Cities Conference in Bonn: 3rd Global Forum on Urban Resilience and Adaptation, 14 May 2012, Bonn, Germany.

Hausler, E., 2010. Building earthquake-resistant houses in Haiti: the homeowner-driven model. *Innovations*, 5 (4), 91–115.

Hills, D., and Bennett, A., 2010. *Framework for developing climate change adaptation strategies and action plans for agriculture in Western Australia*. Perth: Department of Agriculture and Food, Government of Western Australia.

Holden, S., 2004. *Mainstreaming HIV/AIDS in development and humanitarian programmes*. Oxford: Oxfam GB.

Holling, C., 1973. Resilience and stability of ecological systems. *Annual Review of Ecology and Systematics*, 4, 1–23.

IDS (Institute for Development Studies), 2012. *Adaptive social protection: making concepts a reality – guidance notes for practitioners*. Brighton, UK: IDS.

IFRC (International Federation of the Red Cross), 2004. *World disasters report: from risk to resilience – helping communities cope with crisis*. Geneva: IFRC.

IFRC (International Federation of the Red Cross), 2007. *How to do a VCA: a practical step-by-step guide for Red Cross Red Crescent staff and volunteers*. Geneva: IFRC.

IFRC (International Federation of the Red Cross), 2010. *World disasters report: urban risk*. Geneva: IFRC.

IISD (International Institute for Sustainable Development), 2007. *CRiSTAL Community-based risk screening – adaptation and livelihoods user's manual. A decision support tool for assessing and enhancing project impacts on local adaptive capacity to climate variability and climate change. Version 3.0*. IUCN, IISDF, SEI, InterCooperation.

IOI (International Ocean Institute), 2012. *Sharing risks, vulnerabilities and smart growth challenges of global cities in the context of global change*. Presented at the Coastal Cities Summit: Sharing Solutions for Success, 30 April–3 May 2012, St Petersburg, Florida.

IPCC (Intergovernmental Panel on Climate Change), 1997. *IPCC Adaptation Experts' Meeting Report*. 20–22 March 1997, Amsterdam, the Netherlands.

IPCC (Intergovernmental Panel on Climate Change), 2000. *Methodological and technological issues in technology transfer*. A special report of the Intergovernmental Panel on Climate Change [B. Metz *et al.*, eds.]. Cambridge: Cambridge University Press.

IPCC (Intergovernmental Panel on Climate Change), 2001. *Climate change 2001: impacts, adaptation and vulnerability*. Contribution of Working Group II to the

Third Assessment Report of the Intergovernmental Panel on Climate Change [J. McCarty *et al.*, eds.]. Cambridge: Cambridge University Press.

IPCC (Intergovernmental Panel on Climate Change), 2007a. *Climate change 2007: impacts, adaptation and vulnerability.* Contribution of Working Group II to the Fourth Assessment Report of the Intergovernmental Panel on Climate Change [M. Parry *et al.*, eds.]. Cambridge: Cambridge University Press.

IPCC (Intergovernmental Panel on Climate Change), 2007b. Annex II: Glossary. *In: Climate change 2007: Synthesis report.* Contribution of Working Groups I, II and III to the Fourth Assessment Report of the Intergovernmental Panel on Climate Change. Geneva: IPCC, 76–89.

IPCC (Intergovernmental Panel on Climate Change), 2007c. *Climate change 2007: the physical science basis.* Contribution of Working Group I to the Fourth Assessment Report of the Intergovernmental Panel on Climate Change [S. Solomon *et al.*, eds.]. Cambridge: Cambridge University Press.

IPCC (Intergovernmental Panel on Climate Change), 2012a. *Managing the risks of extreme events and disasters to advance climate change adaptation (SREX).* A Special Report of Working Groups I and II of the Intergovernmental Panel on Climate Change [C. Field *et al.*, eds.]. Cambridge: Cambridge University Press.

IPCC (Intergovernmental Panel on Climate Change), 2012b. Glossary of terms. *In: Managing the risks of extreme events and disasters to advance climate change adaptation (SREX).* A Special Report of Working Groups I and II of the Intergovernmental Panel on Climate Change [C. Field *et al.*, eds.]. Cambridge: Cambridge University Press, 555–564.

IRP (International Recovery Platform), 2010a. *Guidance note on recovery: livelihoods.* Geneva: IRP, UNISDR, UNDP.

IRP (International Recovery Platform), 2010b. *Guidance note on recovery: climate change.* Geneva: IRP, UNISDR, UNDP.

ISET (Institute for Social and Environmental Transition), 2010. *The shared learning dialogue: building stakeholder capacity and engagement for resilience action.* Boulder, CO: ISET.

ISO (International Standards Office), 2009. *ISO 31000 Risk management: principles and guidelines.* Geneva: ISO.

Jabeen, H., Johnson, C., and Allen, A., 2010. Built-in resilience: learning from grassroots coping strategies for climate variability. *Environment and Urbanization*, 22, 415–431.

Jacobs, K., and Williams, S., 2011. What to do now? Tensions and dilemmas in responding to natural disasters: a study of three Australian state housing authorities. *International Journal of Housing Policy*, 112, 175–193.

Jeffrey, S., 2011. Ushahidi: crowdmapping collective that exposed Kenyan election killings – how coders built a platform for justice and accountability using mobile phones, text messages and a Google map. *The Guardian NewsBlog*, 7 April 2011.

Jerneck, A., *et al.*, 2011. Structuring sustainability science. *Sustainability Science*, 6, 69–82.

Jha, A. [Alok], 2006. Boiled alive. *The Guardian*, 26 July 2006.

Jha, A. [Abhas], Bloch, R., and Lamond, J., 2012. *Cities and flooding: a guide to integrated urban flood risk management for the 21st century.* Washington, DC: The World Bank.

Johnson, H., and Wilson, G., 2006. North–south/south–north partnerships: closing the 'mutuality gap'. *Public Administration and Development*, 26 (1), 71–80.

Jonsson, A., 2012. *Toolbox for climate change adaptation.* Lecture for the course 'MVENI18 Klimatstrategiska metoder' ['Methods for climate change mitigation and adaptation'] at Lund University. 21 May 2012, Lund, Sweden.

Jonsson, A., *et al.*, 2011. *Verktygslåda för klimatanpassningsprocesser – från sårbarhetsbedömning till sårbarhetshantering [Toolbox for climate change adaptation – from vulnerability assessments to vulnerability management.]* (CSPR No. 11:01). Norrköping: Centre for Climate Science and Policy Research.

Jonsson, A., *et al.*, 2010. *Institutional aspects of adaptation – participatory approaches in Nordic cities and sectors.* Presented at the Climate Adaptation in the Nordic Countries: Science, Practice, Policy, Stockholm University, 8–11 November 2010, Stockholm, Sweden.

Jürgens, I., 2011. Seminar on 'Klimatanpassning i Europa – strategi i utveckling' ['Climate change adaptation in Europe – a strategy under development'], 14 December 2011, Lund, Sweden.

Kates, R., *et al.*, 2001. Sustainability science. *Science*, 292, 641–642.

Kates, R., Travis, W., and Wilbanks, T., 2012. Transformational adaptation when incremental adaptations to climate change are insufficient. *Proceedings of the National Academy of Sciences*, 109 (19), 7156–7161.

Kay, M., 1982. *Energy efficient site planning handbook.* Sydney: Housing Commission of New South Wales.

Kazmierczak, A., and Carter, J., 2010. *Adaptation to climate change using green and blue infrastructure: a database of case studies.* Manchester, UK: University of Manchester.

Khan, A., 2008. Earthquake safe traditional house construction practices in Kashmir: State of Jammu & Kashmir, Northern India. *In*: R. Shaw, N. Uy and J. Baumwoll, eds. *Good practices and lessons learned from experiences in the Asia–Pacific region.* Bangkok: UNISDR.

Kim, H., 1992. Urban heat island. *International Journal of Remote Sensing*, 13 (12), 2319–2336.

Klein, R., and MacIver, D., 1999. Adaptation to climate variability and change: methodological. *Mitigation and Adaptation Strategies for Global Change*, 4, 189–198.

Kleinfield, N., 2011. How hot is 104? New York counts the miseries. *The New York Times*, 23 July 2011, A1.

Klinenberg, E., 2003. *Heat wave: a social autopsy of disaster in Chicago.* Chicago, IL: University of Chicago Press.

Koch-Nielsen, H., 2002. *Stay cool: a design guide for the built environment in hot climates.* London and New York: Routledge.

Konya, A., and Swanepoel, C., 1980. *Design primer for hot climates.* London: Architectural Press.

Krantz, L., 2001. *The sustainable livelihood approach to poverty reduction: an introduction.* Stockholm: Swedish International Development Cooperation Agency (SIDA).

La Trobe, S., and Davis, I., 2005. *Mainstreaming disaster risk reduction: a tool for development organisations.* Teddington, UK: Tearfund.

Lang, D., *et al.*, 2012. Transdisciplinary research in sustainability science: practice, principles, and challenges. *Sustainability Science*, 7, 25–43 [special issue].

Larsen, J., 2006. *Setting the record straight: more than 52,000 Europeans died from heat in summer 2003.* Washington, DC: Earth Policy Institute.

LeGates, R., and Stout, F., 2000. *The city reader.* 2nd ed. London and New York: Routledge.

Leonardsen, L., 2012. *Reality check: adaptation on the ground, Copenhagen, Denmark Workshop.* Presented at the Resilient Cities Conference, 14 May 2012, Bonn, Germany.

Lindblad, J., 2012. Analysing citizens' adaptive capacity: individual adaptation strategies in Helsingborg, Sweden before, during and after the advent storm in 2011. Thesis (Masters). Lund University.

Loftus, A.-C., *et al.*, 2011. *Adapting urban water systems to climate change: a handbook for decision makers at the local level.* Freiburg: ICLEI European Secretariat.

Lomma Kommun, 2012. *Planbeskrivning: Detaljplan för del av Önnerup 1:2 m. fl. Område mellan Bjärreds tätort och Haboljung, Lomma kommun, Skåne län [Plan description: development plan for part of Önnerup 1:2 etc. Area between the conurbation of Bjärred and Haboljund, Lomma municipality, Skåne region].* Lomma: Miljö-och byggförvaltningen i Lomma kommun.

Lonergan, S., *et al.*, 1999. *Global environmental change and human security: GECHS science plan.* Report No. 11. Bonn: International Human Dimensions Programme on GEC (IHDP).

Lourenço, T., *et al.*, 2009. *Outcomes of the 1st international CIRCLE workshop on climate change adaptation.* Budapest: CIRCLE ERA-Net.

Lutz, W., 2010. *Improving education as key to enhancing adaptive capacity in developing countries.* Presented at the International Workshop on 'The Social Dimension of Adaptation to Climate Change', organized by the International Center for Climate Governance (ICCG), 18–19 February 2010, Venice, Italy.

McAdoo, B., Moore, A., and Baumwoll, J., 2009. Indigenous knowledge and the near field population. *Natural Hazards*, 48, 73–82.

McCormick, K., *et al.*, 2013. Advancing sustainable urban transformation [editorial]. *Journal of Cleaner Production*, 50, 1–11 [special issue on 'Advancing sustainable urban transformation'].

McGranahan, G., Balk, D., and Anderson, B., 2007. The rising tide: assessing the risks of climate change and human settlements in low-elevation coastal zones. *Environment and Urbanization*, 19 (1), 17–37.

McGranahan, G., Satterthwaite, D., and Tacoli, C., 2004. *Rural–urban change, boundary problems and environmental burdens.* London: IIED.

Marulanda, M., *et al.*, 2010. *Design and implementation of seismic risk insurance to cover low-income homeowners by a cross-subsidy strategy.* Presented at the 14th European Conference on Earthquake Engineering, Ohrid, Macedonia.

Massey, E., and Bergsma, E., 2008. *Assessing adaptation in 29 European countries.* Amsterdam: IVM.

Matthew, R., *et al.*, 2010. *Global environmental change and human security.* Cambridge, MA: MIT Press.

Matthies, F., *et al.*, 2008. *Heat-health action plans – guidance.* Copenhagen: WHO Regional Office for Europe.

Mechler, R., and Bayer, J., 2006. *Disaster insurance for the poor? A review of microinsurance for natural disaster risks in developing countries.* Laxenberg, Austria: ProVention/IIASA.

Meister, H., *et al.*, 2009. *Floating houses and mosquito nets: emerging climate change adaptation strategies around the world.* IFOK Study. Boston: Meister Consultants Group, Inc.

Mercer, J., 2010. Disaster risk reduction or climate change adaptation: are we reinventing the wheel? *Journal of International Development*, 22, 247–264.

Mickwitz, P., *et al.*, 2009. *Climate policy integration, coherence and governance.* PEER Report No 2. Helsinki: Partnership for European Environmental Research.

Mideksa, T., and Kalbekken, S., 2010. The impact of climate change on the electricity market: a review. *Energy Policy*, 38 (7), 3579–3585.

Mitchell, T., 2003. *An operational framework for mainstreaming disaster risk reduction.* Disaster Studies Working Paper 8. Benfield Hazard Research Centre, London.

Molin Valdes, H., Amaratunga, D., and Haigh, R., 2013. Special issue on 'Making cities resilient: from awareness to implementation'. *International Journal of Disaster Resilience in the Built Environment*, 4 (1).

Morduch, J., 1999. Between the state and the market: can informal insurance patch the safety net? *The World Bank Research Observer*, 14 (2), 187–207.

Morss, R., *et al.*, 2011. Improving societal outcomes of extreme weather in a changing climate: an integrated perspective. *Annual Review of Environment and Resources*, 36, 1–25.

Moser, C., and Satterthwaite, D., 2008. *Towards pro-poor adaptation to climate change in the urban centres of low- and middle-income countries.* Human Settlements Discussion Paper Series: Climate Change and Cities – 3. London: IIED.

MSB (Swedish Civil Contingency Agency), 2012a. *Risk-och sårbarhetsanalyser 2011: Återkoppling av landstingens risk- och sårbarhetsanalyser [Risk and vulnerability analyses 2011: feedback from the risk and vulnerability analyses of the county councils].* Stockholm: MSB.

MSB (Swedish Civil Contingency Agency), 2012b. *Översvämningar i Sverige 1901–2010 [Floods in Sweden 1901–2010].* Stockholm: MSB.

Mukheibir, P., and Ziervogel, G., 2006. *Framework for adaptation to climate change in the City of Cape Town (FAC4T).* Cape Town: Environmental Planning Department.

Mumford, L., 1968. *The city in history: its origins, its transformations, and its prospects.* New York: Harcourt, Brace & World.

Munich Re Group, 2004. *Megacities; megarisks; trends and challenges for insurance and risk management.* Munich: Munich Reinsurance.

Naone, E., 2011. Internet activists mobilize for Japan: programmers and volunteers collect crucial aid information and raise relief funds. 14 March 2011, *Technology Review.* Cambridge, MA: MIT.

NYCDOB (New York City Department of Buildings), 2011. *Request for proposals for the 2011 NYC CoolRoofs Initiative.* New York: NYCDOB.

O Globo, 2013. Bairro Carioca, em Triagem, já sofre com inundações [Bairro Carioca, in Triagem, already suffers from floods]. *O Globo,* 21 January 2013.

O'Brien, K., 2011. Global environmental change II: from adaptation to deliberate transformation. *Progress in Human Geography,* 35 (5), 667–676.

O'Brien, K., and Leichenko, R., 2008. *Environmental change and globalization: double exposures.* Oxford: Oxford University Press.

O'Brien, K., *et al.,* 2007. Why different interpretations of vulnerability matter in climate change discourses. *Climate Policy,* 7 (1), 73–88.

O'Meara, M., 1999. *Reinventing cities for people and the planet.* Washington, DC: Worldwatch Institute.

Olgay, V., 1963. *Design with climate: bioclimatic approach to architectural regionalism.* Princeton, NJ: Princeton University Press.

Olukoya, S., 2011. Nigeria's plastic bottle house. *BBC Africa,* 9 November 2011.

Oxfam, 2002. *PCVA Participatory capacities and vulnerabilities assessment: finding the link between disasters and development.* Oxford: Oxfam GB.

Panda, G.R., 2011. Delivering promises: budgeting for adaptation to climate change. *India Economy Review,* 8, 46–54.

PDC (Pacific Disaster Center), 2005. *Kathmandu Valley, Nepal: disaster risk management profile.* 3CD City Profiles Series. PDC and Earthquake and Megacities Initiative (EMI).

Pelling, M., 2003. *The vulnerability of cities: natural disasters and social resilience.* London: Earthscan.

Pelling, M., 2007. *Urbanization and disaster risk.* Panel contribution to the Population–Environment Research Network Cyberseminar on Population and Natural Hazards, 5–19 November 2007.

Pelling, M., 2011. Urban governance and disaster risk reduction in the Caribbean: the experiences of Oxfam GB. *Environment and Urbanization,* 23 (2), 383–400.

Pelling, M., 2012. *Adaptation to climate change: from resilience to transformation.* Abingdon and New York: Routledge.

Pelling, M., and Manuel-Navarrete, D., 2011. From resilience to transformation: the adaptive cycle in two Mexican urban centers. *Ecology and Society,* 16 (2), 11.

Pelling, M., and Wisner, B., 2009. *Disaster risk reduction: cases from urban Africa.* London: Earthscan.

Pelling, M., *et al.,* 2008. Shadow spaces for social learning: a relational understanding of adaptive capacity to climate change within organisations. *Environment and Planning A,* 40 (4), 867–884.

Pizarro, R., 2009. Urban form and climate change: towards appropriate development patterns to mitigate and adapt to global warming. *In*: S. Davoudi, J. Crawford and A. Mehmood, eds. *Planning for climate change: strategies for mitigation and adaptation for spatial planners.* London: Earthscan, 33–45.

Plan International, 2012. *Climate extreme: how young people can respond to disasters in a changing world.* Woking, UK: Plan International and Children in a Changing Climate.

Practical Action, 2010. *Practice briefing – integrating approaches: sustainable livelihoods, disaster risk reduction and climate change adaptation.* Rugby, UK: Practical Action.

Rauhala, J., and Schultz, D., 2009. Severe thunderstorm and tornado warnings in Europe. *Atmospheric Research*, 93, 1–3.

Reale, A., 2010. Acts of God(s): the role of religion in disaster risk reduction. *Humanitarian Exchange Magazine*, Issue 48.

Redclift, M., Manuel-Navarrete, D., and Pelling, M., 2011. *Climate change and human security: the challenge to local governance under rapid coastal urbanization.* Cheltenham, UK: Edward Elgar Publishing.

RedDesastres, n.d.. *Manual: tools and lessons learned in Central America. El Salvador.* RedDesastres.

Renn, O., Klinke, A., and Asselt, M., 2011. Coping with complexity, uncertainty and ambiguity in risk governance: a synthesis. *Ambio*, 40 (2), 231–246.

RGS (Royal Geographical Society), 2010. Interview: Prof. Sue Grimmond answers questions on urban climatology. 18 February 2010, *Royal Geographical Society: geography in the news – ask the experts.*

Rhodes, R., 1996. The new governance: governing without government. *Political Studies*, 44, 652–667.

Ribeiro, M., *et al.*, 2009. *Design of guidelines for the elaboration of regional climate change adaptations strategies.* Study for European Commission – DG Environment. Vienna: Ecologic Institute.

Risbey, J., *et al.*, 1999. Scale, context, and decision making in agricultural adaptation to climate variability and change. *Mitigation and Adaptation Strategies for Global Change*, 4 (2), 137–165.

Roaf, S., Crichton, D., and Nicol, F., 2005. *Adapting buildings and cities for climate change.* Amsterdam: Elsevier.

Roberts, D., 2008. Thinking globally, acting locally – institutionalizing climate change at the local government level in Durban, South Africa. *Environment and Urbanization*, 20 (2), 521–537.

Roberts, D., 2010. Prioritizing climate change adaptation and local level resilience in Durban, South Africa. *Environment and Urbanization*, 22 (2), 397–413.

Roberts, D., *et al.*, 2012. Exploring ecosystem-based adaptation in Durban, South Africa: 'learning-by-doing' at the local government coal face. *Environment and Urbanization*, 24 (1), 167–195.

Roberts, P., Ravetz, J., and George, C., 2009. *Environment and the city.* Abingdon and New York: Routledge.

Rode, P., *et al.*, 2012. *Going green: how cities are leading the next economy.* London: LSE Cities.

Roggema, R., 2009. *Adaptation to climate change: a spatial challenge.* New York: Springer.

Rowswell, P., and Clover, J., 2012. *Community based adaptation and mitigation in Southern Africa.* Presented at the Resilient Cities Conference 2012, Session D1 on 'Mitigating and adapting from the bottom up: community-based solutions', 14 May 2012, Bonn, Germany.

SAARC, 2008. *Indigenous knowledge for disaster risk reduction in South Asia.* New Delhi: SAARC Disaster Management Centre.

Salmon, C., 1999. *Architectural design for tropical regions.* New York: John Wiley & Sons.

Sanderson, D., 2000. Cities, disasters and livelihoods. *Environment and Urbanization,* 12 (2), 93–102.

Satterthwaite, D., 2007. Climate change and urbanization: effects and implications for urban governance. *In*: UNDESA, ed. *Population distribution, urbanization, internal migration and development: an international perspective.* New York: UNDESA.

Satterthwaite, D., 2008. Cities' contribution to global warming: notes on the allocation of greenhouse gas emissions. *Environment and Urbanization,* 20, 539–549.

Satterthwaite, D., 2011a. *What role for low-income communities in urban areas in disaster risk reduction?* Background paper prepared for the 'Global Assessment Report on Disaster Risk Reduction (GAR 2011)'. Geneva: UNISDR.

Satterthwaite, D., ed., 2011b. Community-driven disaster risk reduction and climate change adaptation in urban areas [special issue]. *Environment and Urbanization* 23 (2).

Satterthwaite, D., *et al.* eds., 2007a. Special issue on 'Reducing risks to cities from disasters and climate change'. *Environment and Urbanization* 19 (1).

Satterthwaite, D., *et al.*, 2007b. *Adapting to climate change in urban areas: the possibilities and constraints in low- and middle-income nations.* Human Settlements Discussion Paper Series. Theme: Climate change and cities (1). London: IIED.

Schimanski, F., 2008. Stadsbornas kolonilotter livsviktiga vid kriser [City dwellers' allotment gardens: vital in times of crisis]. *Populär Historia.* No. 8/2008, 28 July 2008.

Schinkel, U., Ánh, L., and Schwartze, F., 2011. *How to respond to climate change impacts in urban areas: a handbook for community action.* Brandenburg: Department for Urban Planning and Spatial Design, Brandenburg University of Technology Cottbus.

Schuster, P., 2008. *Klimaanpassungsstrategien in europäischen Nachbarländern – ein Werkstattbericht – Anpassungsstrategien an den Klimawandel in EU27 und ihre potentielle Bedeutung für die Raumentwicklung in Deutschland [Climate change adaptation strategies in European neighbouring countries – a project report – Strategies for climate change adaptation in EU27 and its potential role for the spatial planning and development in Germany].* Bonn: BBR.

Scoones, I., 1998. *Sustainable rural livelihoods: a framework for analysis.* Brighton, UK: IDS.

SDC (Swiss Agency for Development and Cooperation), 2009. *Putting a livelihood perspective into practice: systemic approach to rural development (SARD).* Bern: SDC.

Shah, F., and Ranghieri, F., 2012. *A workbook on planning for urban resilience in the face of disasters: adapting experiences from Vietnam's cities to other cities.* Washington, DC: The World Bank.

Shaw, R., and Sharma, A., eds., 2011. *Climate and disaster resilience in cities.* Community, Environment and Disaster Risk Management Vol. 6. Bingley, UK: Emerald Group Publishing Limited.

Shaw, R., Colley, M., and Connell, R., 2007. *Climate change adaptation by design: a guide for sustainable communities.* London: Town and Country Planning Association (TCPA).

Shaw, R., *et al.*, 2008. *Policy note: indigenous knowledge – disaster risk reduction.* Kyoto: EU, UNISDR, Kyoto University, Seeds.

Simatele, D., 2010. *Climate change adaptation in Lusaka, Zambia: a case study of Kalingalinga and Linda Compounds.* Global Urban Research Centre (GURC) Working Paper No. 6. Manchester: GURC.

Simmel, G., 1960. The metropolis and mental life. *In*: W. Mills, ed. *Images of man: the classic tradition in sociological thinking.* New York: G. Braziller.

Simon, D., 2013. The environmental determinants of human security in the context of climate change. *In*: M. Redclift and M. Grasso, eds. *Handbook of climate change and human security.* Aldershot, UK: Edward Elgar (in press).

Simon, D., and Leck, H., 2010. Urbanizing the global environmental change and human security agendas. *Climate and Development*, 2 (3), 263–275.

Singh, D., 2011. The wave that eats people: the value of indigenous knowledge for disaster risk reduction. 9 August 2011, United Nations Office for Disaster Risk Reduction – *News Archive*.

SKL (Swedish Association of Local Authorities and Regions), 2011. *SKL granskar kommunernas arbete med klimatanpassning [SKL reviews the municipalities' work with climate change adaptation].* Stockholm: SKL.

Smit, B., and Wandel, J., 2006. Adaptation, adaptive capacity and vulnerability. *Global Environmental Change*, 16, 282–292.

Smit, B., *et al.*, 2001. Adaptation to climate change in the context of sustainable development and equity. *In*: J. McCarthy *et al.*, eds. *Climate change 2001: impacts, adaptation, and vulnerability.* New York: Cambridge University Press, 877–912.

Soltesova, K., *et al.*, 2012. Community participation in urban adaptation to climate change: potential and limits for CBA approaches. *In*: J. Ayers *et al.*, eds., *Community-based adaptation: scaling it up.* London: Earthscan.

Steffen, W., *et al.*, 2004. *Global change and the Earth system: a planet under pressure.* New York: Springer.

Stein, A., 2010. Urban poverty, social exclusion and social housing finance: the case of PRODEL in Nicaragua. Thesis (Doctorate). Lund University.

Stephenson, A., *et al.*, 2010. *Benchmark resilience: a study of the resilience of organisations in the Auckland Region*. Canterbury: Resilient Organisations.

Stern, N., 2006. *Stern review on the economics of climate change*. London: UK Government Economic Service.

Stive, M., 2012. Lecture on 'Rising tides – the Dutch strategy on how to cope with climate change in the 21st century'. 24 May 2012, Lund, Sweden.

Striessnig, E., Lutz, W., and Patt, A., 2013. Effects of educational attainment on climate risk vulnerability. *Ecology and Society*, 18 (1), 16.

SVS (Secretaria de Vigilância em Saúde), 2011. *Saiba como agir em caso de enchentes [Know how to act in case of floods]*. Ministério da Saúde (Brazilian Ministry of Health).

Swart, R., *et al.*, 2010. Europe adapts to climate change: comparing national adaptation strategies. *Global Environmental Change*, 20 (3), 440–450.

Tacoli, C., 2003. The links between urban and rural development. *Environment and Urbanization*, 15 (1), 3–12.

Tacoli, C., 2012. *Urbanization, gender and urban poverty: paid work and unpaid carework in the city*. Urbanization and emerging population issues. Working Paper 7. London and New York: IIED and UNFPA.

Taylor, A., Thorne, S., and Mquad, L., 2008. Technologies for adaptation. *Tiempo: A Bulletin on Climate and Development*, 67, 8–13.

Tearfund, 2009. *CEDRA Climate change and environmental degradation risk and adaptation assessment*. Teddington, UK: Tearfund.

Tearfund, 2011. *Reducing risk of disaster in our communities*. ROOTS. 2nd ed. Teddington, UK: Tearfund.

Tehler, H., 2012. *Risk governance: understanding the management of risks in cities and regions. Training Regions Professional Papers*, 1 (2). Lund University.

Tehler, H., 2013. *A general framework for risk assessment V 1.3*. Educational training material. Lund: Department of Fire Safety Engineering and Systems Safety, Lund University.

Thompson, M., 2011. The quest for 'clumsy solutions' in Nepal's mountains. *Options*, Winter 2011/12, 12–13.

Thompson, M., and Wildavsky, A., 1986. A poverty of distinction: from economic homogeneity to cultural heterogeneity in the classification of poor people. *Policy Science*, 19, 163–199.

Thompson, M., Ellis, R., and Wildavsky, A., 1990. *Cultural theory*. Boulder, CO: Westview Press.

Thompson, M., Warburton, M., and Hatley, T., 2007. *Uncertainty: on a Himalayan scale – an institutional theory of environmental perception and a strategic framework for the sustainable development of the Himalaya*. Patan Dhoka: Himal Books.

Tjiputra, J., and Otterå, O., 2011. Role of volcanic forcing on future global carbon cycle. *Earth System Dynamics*, 2, 53–67.

Troedsson, T., 2011. *Utan flöde – inget kul [No flow – no fun]*. Presenation at the Training Regions resilient urban flow training workshop, 13 October 2011, Malmö, Sweden.

Tubby, K., and Webber, J., 2010. Pests and diseases threatening urban trees under a changing climate. *Forestry*, 83 (4), 451–459.

Tudehope, M., 2011. 'Bat people' and 'floating houses': hope in the lowliest of Manila's slums. *The Global Urbanist: Communities*, 29 March 2011.

Turpeinen, H., *et al.*, 2008. Effect of ice sheet growth and melting on the slip evolution of thrust faults. *Earth and Planetary Science Letters*, 269 (1–2), 230–241.

Twigg, J., 2004. *Disaster risk reduction: mitigation and preparedness in development and emergency programming*, No. 9. Good Practice Review. London: ODI.

UKCIP (UK Climate Impacts Programme), 2007. *Identifying adaptation options*. Oxford: UKCIP.

UN Water, 2007. *World Water Day 2007: coping with water scarcity: every drop counts*. Rome: UN Water, United Nations.

UN-IATF/DR., 2006. *On better terms: a glance at key climate change and disaster risk reduction concepts*. Geneva: United Nations Working Group on Climate Change and Disaster Risk Reduction of the Inter-Agency Task Force on Disaster Reduction (UN-IATF/DR).

UNAIDS (Joint United Nations Programme on HIV/AIDS), 2008. *Climate change and AIDS: a joint working paper*. Nairobi: UNAIDS.

UNCSD (United Nations Conference on Sustainable Development), 2012. *RIO 2012 issues briefs* (No. 5). Rio de Janeiro: UNCSD Secretariat.

UNDP (United Nations Development Programme), 2004. *Reducing disaster risk: a challenge for development – a global report*. Geneva: Bureau for Crisis Prevention and Recovery, UNDP.

UNDP (United Nations Development Programme), 2011. *Disaster–conflict interface: comparative experiences*. Geneva: Bureau for Crisis Prevention and Recovery, UNDP.

UNDP/CWGER (United Nations Development Programme/Cluster Working Group on Early Recovery), 2008. *Guidance note on early recovery*. Geneva: UNDP and CWGER.

UNESCO (United Nations Educational Scientific and Cultural Organization), 2012. *A guide on adaptation options for local decision-makers: guidance for decision making to cope with coastal changes in West Africa*. Paris: UNESCO.

UNFCCC (United Nations Framework Convention on Climate Change), 2007. *Climate change: impacts, vulnerabilities and adaptation in developing countries*. Bonn: UNFCCC.

UNHABITAT (United Nations Human Settlements Programme), 2003. *The challenge of slums: global report on human settlements 2003*. London: Earthscan.

UNHABITAT (United Nations Human Settlements Programme), 2006. *State of the world's cities 2006/2007: the millennium development goals and urban sustainability –30 years of shaping the Habitat agenda*. London: Earthscan.

UNHABITAT (United Nations Human Settlements Programme), 2007. *Enhancing urban safety and security: global report on human settlements 2007*. London: Earthscan.

UNHABITAT (United Nations Human Settlements Programme), 2008. *State of the world's cities 2008/2009: harmonious cities.* London: Earthscan.

UNHABITAT (United Nations Human Settlements Programme), 2009. *Planning sustainable cities: global report on human settlements 2009.* London: Earthscan.

UNHABITAT (United Nations Human Settlements Programme), 2010a. *State of the world's cities 2010/2011: bridging the urban divide.* London: Earthscan.

UNHABITAT (United Nations Human Settlements Programme), 2010b. *Gender equality for smarter cities: challenges and progress.* Nairobi: UNHABITAT.

UNHABITAT (United Nations Human Settlements Programme), 2011. *Cities and climate change: global report on human settlements 2011.* London: Earthscan.

UNISDR (United Nations Office for Disaster Risk Reduction), 2004. *Living with risk: a global review of disaster reduction initiatives.* Geneva and New York: UNISDR.

UNISDR (United Nations Office for Disaster Risk Reduction), 2005. *Hyogo Framework for Action 2005–2015: building the resilience of nations and communities to disaster.* 22 January 2005, Kobe, Kyogo, Japan: UNISDR.

UNISDR (United Nations Office for Disaster Risk Reduction), 2008a. *Briefing note 01: climate change and disaster risk reduction.* Geneva: UNISDR.

UNISDR (United Nations Office for Disaster Risk Reduction), 2008b. *Links between disaster risk reduction, development and climate change. A briefing for Sweden's commission on climate change and development.* Geneva: UNISDR.

UNISDR (United Nations Office for Disaster Risk Reduction), 2009. *Terminology: disaster risk reduction.* Geneva: UNISDR.

UNISDR (United Nations Office for Disaster Risk Reduction), 2010a. *Making Cities Resilient: my city is getting ready* (Campaign kit for the 2012–2011 World Disaster Reduction Campaign). Geneva: UNISDR.

UNISDR (United Nations Office for Disaster Risk Reduction), 2010b. *Local governments and disaster risk reduction: good practices and lessons learned. A contribution to the 'Making Cities Resilient' Campaign.* Geneva: UNISDR.

UNISDR (United Nations Office for Disaster Risk Reduction), 2010c. *Briefing note 02: adaptation to climate change by reducing disaster risks – country practices and lessons.* Geneva: UNISDR.

UNISDR (United Nations Office for Disaster Risk Reduction), 2011a. *Climate change adaptation and disaster risk reduction in Europe: a review of risk governance.* UNISDR, EUR-OPA, Council of Europe.

UNISDR (United Nations Office for Disaster Risk Reduction), 2011b. *2011 global assessment report on disaster risk reduction: revealing risk, redefining development.* Geneva: UNISDR.

UNISDR (United Nations Office for Disaster Risk Reduction), 2012a. *2011 – disasters in numbers.* UNISDR, USAID, CRED.

UNISDR (United Nations Office for Disaster Risk Reduction), 2012b. *How to make cities more resilient: a handbook for local government leaders. A contribution to the global campaign 2010–2015 'Making Cities Resilient – my city is getting ready!'* Geneva: UNISDR.

UNISDR (United Nations Office for Disaster Risk Reduction), 2012c. *Making Cities Resilient report 2012: my city is getting ready! A global snapshot of how local governments reduce disaster risk*. First Edition for the 5th World Urban Forum. Geneva: UNISDR.

UNISDR (United Nations Office for Disaster Risk Reduction), 2012d. *Managing disaster risks for resilience in the 21st century. Keynote address by Margareta Wahlström – UN Secretary-General's Special Representative for Disaster Risk Reduction*. Presented at the 4th International Disaster and Risk Conference Integrative Risk Management in a Changing World – Pathways to a Resilient Society, 26 August 2012, Davos, Switzerland.

United Nations, 2012. *Report of the United Nations conference on sustainable development*. New York: United Nations.

USGS (United States Geological Survey), 2012. *Volcano hazards program – volcanic ash: effects and mitigation strategies*. Reston: USGS.

van Wesemael, B., *et al.*, 1998. Collection and storage of runoff from hillslopes in a semi-arid environment: geomorphic and hydrologic aspects of the aljibe system in Almeria Province, Spain. *Journal of Arid Environments*, 40, 1–14.

Wahlström, M., 2012. *Launch of 'How to make cities more resilient: handbook for local government leaders'*. Resilient Cities Conference, 14 May 2012, Bonn, Germany.

Wamsler, C., 2002. Disaster risk management: measures for houses and settlements in risk areas. *TRIALOG (Journal for Planning and Building in the 'Third World')*, 73, 32–35.

Wamsler, C., 2006a. Managing urban disasters [editorial]. *Open House International*, 31 (1), 4–9 [special issue on 'Managing urban disasters'].

Wamsler, C., 2006b. Integrating risk reduction, urban planning and housing: lessons from El Salvador. *Open House International*, 31 (1), 71–83 [special issue on 'Managing urban disasters'].

Wamsler, C., 2006c. Mainstreaming risk reduction in urban planning and housing: a challenge for international aid organisations. *Disasters*, 30 (2), 151–77.

Wamsler, C., 2007a. Bridging the gaps: stakeholder-based strategies for risk reduction and financing for the urban poor. *Environment and Urbanization*, 19 (1), 115–142 [special issue on 'Reducing risks to cities from climate change and disasters'].

Wamsler, C., 2007b. Coping strategies in urban slums, Box 6.5. *In: State of the world 2007: our urban future*. New York: The Worldwatch Institute, Norton & Company, 124.

Wamsler, C., 2007c. Managing urban disaster risk: analysis and adaptation frameworks for integrated settlement development programming for the urban poor. Thesis (Doctorate). Lund University.

Wamsler, C., 2008b. Achieving urban resilience: understanding and tackling disasters from a local perspective. *Urban Design and Planning*, 161 (DP4), 163–171.

Wamsler, C., 2008c. Climate change impacts on cities: ignore, mitigate or adapt? *TRIALOG (Journal for Planning and Building in the 'Third World')*, 97 (2), 4–10.

Wamsler, C., 2008d. Mainstreaming HIV/AIDS in settlement development planning [editorial]. *Open House International*, 33 (4), 6–7 [special issue on 'HIV/AIDS and settlement development planning'].

Wamsler, C., 2009a. *Operational framework for integrating risk reduction and climate change adaptation into urban development*. Brookes World Poverty Institute (BWPI), Working Paper Series No. 101 & Global Urban Research Centre (GURC), Working Paper Series No. 3. Manchester, UK: BWPI/GURC.

Wamsler, C., 2009b. *Urban risk reduction and adaptation: how to promote resilient communities and adapt to increasing disasters and changing climatic conditions?* Saarbrücken: VDM Verlag.

Wamsler, C., and Brink, E., 2014. Adaptive capacity: from coping to sustainable transformation. *In*: S. Eriksen *et al.*, eds. *Climate change adaptation and development: transforming paradigms and practices*. Abingdon and New York: Routledge.

Wamsler, C., and Lawson, N., 2011. The role of formal and informal insurance mechanisms for reducing urban disaster risk: a south–north comparison. *Housing Studies*, 26 (2), 197–223.

Wamsler, C., and Lawson, N., 2012. Complementing institutional with localised strategies for climate change adaptation: a south–north comparison. *Disasters*, 36 (1), 28–53.

Wamsler, C., and Umaña, C., 2003. *El Salvador: proyecto de reconstrucción con inclusión de la gestión de riesgo [Reconstruction programming with integration of disaster risk management]*. Deutsche Gesellschaft für Technische Zusammenarbeit (GTZ) GmbH. Eschborn: GTZ.

Wamsler, C., Brink, E., and Rantala, O., 2012. Climate change, adaptation, and formal education: the role of schooling for increasing societies' adaptive capacities in El Salvador and Brazil. *Ecology and Society*, 17 (2) Art. 2 [special issue on 'Education and differential vulnerability to natural disasters'].

WCED (World Commission on Environment and Development), 1987. *Our common future: report of the World Commission on Environment and Development*. Oxford: Oxford University Press.

Weber, M., 1966. *The city*. 2nd ed. New York: Free Press.

WEF (World Economic Forum), 2012. *Global risks 2012 – seventh edition: an initiative of the Risk Response Network*. Cologny and Geneva: WE Forum.

Weinstein, M., 2010. Sustainability science: the emerging paradigm and the ecology of cities [editorial]. *Sustainability: Science, Practice, & Policy*, 6 (1), 1–5.

West, H., 2006. *Victims of violence in times of disaster or emergency*. Presented at the After the Crisis: Healing from Trauma after Disasters Expert Panel Meeting, 24–25 April 2006, Bethesda.

Wilbanks, T., and Kates, R., 2010. Beyond adapting to climate change: embedding adaptation in responses to multiple threats and stresses. *Annals of the Association of American Geographers*, 100 (4), 719–728.

Wirth, L., 1938. Urbanism as a way of life. *American Journal of Sociology*, 1, 1–24.

Wisner, B., 2003. Disaster risk reduction in megacities: making the most of human and social capital. *In*: A. Kreimer, M. Arnold and A. Carlin, eds. *Building safer cities: the future of disaster risk, disaster risk management series*. Washington, DC: The World Bank.

Wisner, B., *et al.*, 2004. *At risk: natural hazards, people's vulnerability and disasters*. 2nd ed. London and New York: Routledge.

Wisner, B., Gaillard, J.-C., and Kelman, I., 2011. *Handbook of hazards and disaster risk reduction*. Abingdon and New York: Routledge.

World Bank, 2005. *Surviving disasters and supporting recovery: a guidebook for microfinance institutions* (Disaster Risk Management Working Papers Series No. 10). Washington, DC: The World Bank.

World Bank, 2010. *Cities and climate change: an urgent agenda*. Urban Development Series Knowledge Papers. Washington, DC: The World Bank.

World Bank, 2012. *Climate change, disaster risk, and the urban poor: cities building resilience for a changing world*. Washington, DC: The World Bank.

World Bank, 2013. *Guide to climate change adaptation in cities*. Washington, DC: The World Bank.

Worldwatch Institute, 2007. Reducing natural disaster risk in cities. *In*: *State of the world 2007: our urban future*. New York: The Worldwatch Institute, Norton & Company, 112–133.

Yohe, G., *et al.*, 2006. *A synthetic assessment of the global distribution of vulnerability to climate change from the IPCC perspective that reflects exposure and adaptive capacity*. New York: CIESIN, Columbia University.

Yu, J., 2006. *Wind effect on pedestrians*. Hong Kong: Hong Kong Institute of Steel Construction.

Zimov, S.A., Schuur, E.A.G., and Chapin, F.S., 2006. Permafrost and the global carbon budget. *Science*, 312 (5780), 1612–1613.

Zolli, A., and Healy, A.M., 2012. *Resilience: why things bounce back*. 1st ed. New York: Free Press.

Index